George Miller Beard

Eating and drinking

A popular manual of food and diet in health and disease

George Miller Beard

Eating and drinking
A popular manual of food and diet in health and disease

ISBN/EAN: 9783337201296

Printed in Europe, USA, Canada, Australia, Japan

Cover: Foto ©berggeist007 / pixelio.de

More available books at **www.hansebooks.com**

Eating and Drinking;

A POPULAR MANUAL OF

FOOD AND DIET IN HEALTH AND DISEASE.

BY

GEORGE M. BEARD, M.D.

NEW YORK:

G. P. PUTNAM & SONS.

1871.

PREFACE.

THIS volume is designed to be a companion treatise to my work on Stimulants and Narcotics. What that work did for negative, this aims to do for positive food—to give a description and history of the principal articles of ordinary consumption ; to show their relation to each other, and their complex influence on the human system, as modified by race, climate, age, sex, habit; and the progress of civilization ; to expose the widely prevailing errors in regard to this subject, and to point out and enforce those practical rules of dietetics which the experience of mankind has illustrated and confirmed.

The two volumes have a mutual dependence, and to a certain extent supplement each other.

The terms *positive* and *negative* food, which I first suggested a few years since, I have here retained, although they have never been formally presented to the profession.

The distinction which they imply between the respective action of ordinary food and stimulants and narcotics, though not complete or satisfactory, is yet convenient and clear, and in the present state of our ignorance of the subject, is perhaps about as far as we can go. Strictly speaking, we do not, in the last analysis, know how food of any kind sustains the system. Why it is and how it is that beef and bread, fruits and vegetables, when digested and taken into the circulation, supply heat and force and the material for growth and repair, we do not precisely know; and all the labors of chemists and physiologists leave us still in the dark.

To those who have derived their conceptions of dietetics from the noisy and ignorant charlatans who in this country more than in any other, through the unfortunate indifference of the profession, have monopolized the public attention, the facts and conclusions of this work must seem revolutionary, if not incredible.

Those who believe that the chief end of man is to be hungry, will find in this volume little to confirm, perhaps much to disturb that belief. Those who expect that the natural appetite, as a guide in determining the quantity and quality of our food, is to be superseded by the chemist or the scales, and who are unwilling to listen to opposing arguments, will need to read no further than this preface.

The vast army of Jeremiahs who have gone up and down the land, predicting that our gluttony would be our ruin, will be surprised and gratified at the suggestion that the vice of over-eating, once so prevalent, is under civilization fast disappearing, and that it is no longer national or universal; that indeed the tendency among the intellectual and cultivated classes of our time is to eat too little rather than too much.

If there be any who suspect that the classification of Liebig has correctly or exhaustively expressed the chemistry of food, they would do well to consider the very numerous facts which demonstrate that in extremely hot, as well as in extremely cold climates, the so-called heat-producing foods in large quantities are imperatively demanded.

The dogma that fish diet increases the power and activity of the intellect—which, unlike most of the popular errors, has been sanctioned by eminent names in science—which has hitherto never been formally disputed, I have shown to be quite the reverse of the truth, and if there are any who have failed to derive the promised increase of brain-power from a diet composed largely of fish, they will find in this work the consolation that they are not alone in their failure.

The cruel edicts which dietarians have so rigidly enforced —that only one or two varieties should be taken at a time; that there should be no drinking at meals; that the appetite is to be subdued rather than guided, and that a sense of satiety is a conviction of sin—are shown to have no scientific foundation. For those who are proclaiming that the hope of salvation of the race depends on our returning to the dietetic habits of our fathers, I have adduced facts and reasonings, which show that the law of evolution applies to diet as demonstrably as to all else in nature, and that to revive the eating and drinking customs of the past, while retaining our present customs and constitutions, would be to imperil our civilization.

The dietetic treatment which I have recommended for nervous dyspepsia—a disease with which nearly all intelligent Americans become acquainted before they die—will be found to be as much more successful than the opposite method, which has so long prevailed, as it is more scientific and agreeable.

The task which I have here begun of raising hygiene to a science, based on experience, and confirmed, guided and illustrated by chemistry, physiology, and other allied sciences, is one of appalling difficulties, not the least of which is, that first of all it becomes necessary to clear away the rubbish that has been accumulating through centuries of ignorance, and to eradicate errors that have been so long rooted in the popular mind, as to have acquired the strength that belongs to possession, and the authority and sacredness that are attributed to age.

Fully recognizing on the part of a popular scientific writer a double fealty, on the one hand to the science that I represent, and on the other to the people for whom I write, I have sought to make these works as true to scientific methods as though they were designed for an audience strictly professional, and at the same time fully comprehensible to the readers of the ordinary newspaper, which in this country constitutes the almost exclusive intellectual entertainment of those who are able to read at all. It has been my aim to treat the subjects here discussed so thoroughly and so impartially, that even those who question the soundness of my views, may find in these volumes the best materials to sustain their queries, and the strongest weapons with which to combat the doctrines that I teach.

Some of the practical portions of this work—the diet of brain-workers and students—originally appeared in the *Hours at Home* magazine, and the *College Courant*, and it has been a source of gratification to learn that they have been of substantial service to many erring sufferers. My hope is that in this permanent form they may be yet more widely useful.

I desire to express my obligations to Mr. S. B. Noyes, Librarian of the Brooklyn Mercantile Library, who, during all the laborious investigations that the preparation of these works have called forth, has very cheerfully and with remarkable patience, and oftentimes without solicitation, given me the aid of his extensive bibliographical knowledge and experience, and the free range of the remarkably well-selected collection of books under his charge. Similarly, I am indebted to Mr. Hannah, the accomplished Librarian of the Long Island Historical Society.

For information concerning the dietetic habits of the East, I am under obligations to Naotaro Yangimoto, of Japan.

G. M. B.

NEW YORK, *October 1st,* 1871.

SOURCES OF INFORMATION ON THE SUBJECT OF THIS WORK.

UNTIL quite recently, the profession had written scarcely anything of truly real value on the important department of Dietetics.

Within the past few years a number of careful and scholarly observers, especially in Europe, have endeavored to advance and to popularize this branch of hygiene both by laborious experimenting and a judicious use of the pen.

Of the very numerous special writings on this subject that have recently been published in this country, none, so far as I am aware, have been prepared by writers of scientific attainments and reputation ; and most of them contain such a large admixture of error and absurdity that they should be carefully avoided by those who have any regard for scientific truth or for their own physical wellbeing. The latest views of the profession on the subject are found in the following writings, to many of which I have been more or less indebted.

H. LETHEBY.—Lectures on Food, London, 1870.

C. A. CAMERON, M. D.—On Food and Diet, London, 1871.

HORACE DOBELL, M. D.—On Diet and Regimen in Sickness and Health, London, 1870.

WILLIAM BRINTON, M. D.—On Food and its Digestion.

JONATHAN PEREIRA, M. D.—On Food and Diet.

A. BRIGHAM, M. D.—On Mental Exertion in Relation to Health. (This work, now almost forgotten, was certainly far in advance of the age in which it was written, and is well worthy of study. It was but recently that I read it, and found that Dr. Brigham had anticipated me in some views that I have in this work demonstrated by historic facts.)

W. MARCKT.—On the Composition of Food, and how it is adulterated, London, 1866.

THOMAS H. HOSKINS, M. D.—What we Eat, 1861.

AUSTIN FLINT, JR.—On the Physiological Effects of Severe and Protracted Muscular Exercise upon the Excretion of Nitrogen, New York, 1871.

GEO. M. BEARD, M. D.—Our Home Physician, New York, 1869.

LIEBIG.—Animal Chemistry, London, 1840.

HAUGHTON.—Relation of Food to Work, Lancet, Aug. 15, 22, and 29, 1868

PAYEN.—Précis théorique et pratique des substances alimentaires, 1865.

E. LANKESTER.—On Food, 1864.

A. SOYER.—The Pantropheon, or the History of Food and its Preparation from the Earliest Ages of the World, 1853.

Gastronomy, or the School for Good living, London, 1822.

PETER L. SIMMONDS.—The Curiosities of Food, 1857.

AUSTIN FLINT, SR.—Alimentation in Disease. New York Medical Journal, Feb. 1868.

My information concerning the eating customs of different countries was derived from a large variety of works of travel, some of which are included in the following list. For the *generalizations* from the facts I am alone responsible.

P. S. SIMMONDS.—The Commercial Products of the Vegetable Kingdom. London, 1854.

JOHN MONTGOMERY.—Wealth of Nature. Edinburgh, 1870.

J. M. GILLISS.—U. S. Astronomical Expedition. Washington, 1855.

CHARLES DARWIN.—Origin of Species.

————— Varieties of Animals and Plants under domestication. New York, 1868.

————— Descent of Man. New York, 1871.

HERBERT SPENCER.—Principles of Biology. New York, 1867.

HENRY T. BUCKLE.—History of Civilization. New York, 1868.

PROF. AND MRS. LOUIS AGASSIZ.—A Journey in Brazil. New York, 1868.

JAMES RICHARDSON. -Travels in the Great Desert of Sahara, 1848.

HENRY BARTH.—Travels and Discoveries in North and Central Africa. New York, 1857.

H. B. TRISTRAM.—The Great Sahara. London, 1860.

W. WINWOOD READ.- -Savage Africa. New York, 1869.

PAUL B. DU CHAILLU.—Wild Life under the Equator. New York, 1864.

————— Equatorial Africa. New York, 1862.

PAUL B. DU CHAILLU.—A Journey to Ashango. London and New York, 1867.

EMMELINE LOTT.—Harem Life. Philadelphia.

S. S. HILL.—Travels in Peru and Mexico. London, 1860.

M. HUC.—A Journey through Tartary, Thibet and China. New York, 1852.

———— A Journey through the Chinese Empire. New York, 1855.

PLINY MILES.—Rambles in Iceland. New York, 1859.

CHARLES S. FORBES.—Iceland. London, 1860.

JOHN BARROW.—A Visit to Iceland. London, 1835.

ALEXANDER VON HUMBOLDT.—Personal Narrative of Travels. Bohn's Library.
    London, 1852.

N. H. Bishop.—A Thousand Miles Walk across South America. Boston, 1867.

DON RAMON PAEZ.—Wild Scenes in South America. New York, 1862.

PETER SCHMIDTMEYER.—Travels in Chile. London, 1824.

KIDDER AND FLETCHER.—Brazil and the Brazilians.

JOHN CAPPER.—The Three Presidencies of India. London, 1853.

INDIA.—Pictorial, Descriptive and Historical. Bohn's Library. London, 1859.

H. DWIGHT WILLIAMS.—A Year in China. New York, 1869.

WM. M. THOMSON, D.D.—The Land and the Book. New York, 1859.

IDA PFEIFFER.—Travels to Madagascar. New York, 1861.

MARCO POLO.—Travels. Edinburgh, 1844.

F. SHOBERT.—Persia. Philadelphia, 1828.

CUSTINE.—Russia, 1854.

THOMAS FORRESTER.—Norway in 1848 and 1849. London, 1850.

WILLIAM GIFFORD PALGRAVE.—Central and Eastern Arabia. London, 1866.

CHARLES MACFARLANE.—Japan, 1856.

GEORGE KENNAN.—Tent Life in Siberia, 1870.

SAMUEL HAZARD.—Cuba, with pen and pencil, 1871.

ALFRED RUSSELL WALLACE.—The Malay Archipelago, 1869.

———— Travels on the Amazon and Rio Negro. London.

RICHARD J. BUSH.—Reindeer, Dogs and Snow Shoes, 1871.

F. WHYMPER.—Travels and Adventures in Alaska. New York, 1869.

ALBERT S. BICKMORE.—Travels in the Indian Archipelago, 1869.

SPEKE.—Sources of the Nile, 1864.

E. K. KANE.—Arctic Explorations. Philadelphia, 1856.

J. NEVINS.—China and the Chinese. New York, 1869.

J. DOOLITTLE.—Sacred Life of the Chinese. New York, 1865.

VAMBERY.—Travels in Central Asia, 1865.

J. J. HAYES.—The Open Polar Sea, 1867.

CAMPBELL.—Ceylon, 1843.

MANSFIELD PARKYNS.—Life in Abyssinia. New York, 1864.

RICHARD F. BURTON.—The Lake Regions of Central Africa. New York, 1860.

JOHN MIERS.—Travels in Chile and La Plata. London, 1826.

W. S. W. RUSCHENBERGER, M. D.—A Voyage Round the World. Philadelphia, 1838.

JOHN BLACKIE.—Among the Goths and Vandals. London, 1870.

Among the authorities on social customs are the following :

GEO. ROBERTS.—Social History of England. London, 1856.

WILLIAM GOODMAN.—The Social History of Great Britain. New York, 1846.

JAMES LOGAN.—Manners, Customs and Antiquities of Scotland. Hartford, 1855.

ALEXANDER ANDREWS.—The Eighteenth Century. London, 1856.

ROBERT CHAMBERS.—Domestic Annals of Scotland, Edinburgh, 1861.

SHARON TURNER.—History of the Anglo-Saxons. London, 1852.

JOHN DUNLOP.—Artificial and Compulsory Drinking Usages of the United Kingdom. London, 1849.

SAMUEL MOREWOOD.—History of Inebriating Liquors. Dublin, 1838.

GEORGE YOUNG.—A Treatise on Opium. London, 1753.

EDWARD BARRY.—Observations, Historical, Critical and Medical, on the Wines of the Ancients, and the Analogy between them and Modern Wines. London, 1775.

JOHN ELLIS.—Historical Account of Coffee. London, 1764.

J. C. LETTSOM, M. D.—On Tea and Tea Drinking. London, 1762.

Besides these, some of the writings referred to in my work on Stimulants and Narcotics may be consulted with advantage.

CONTENTS.

CHAPTER IV.

CHAPTER V.

CHAPTER VI.

CHAPTER VII.

CHAPTER XIII.

CHAPTER XIV.

" The first requisite for success is to be a good animal."

" Tell me what thou eatest, and I will tell thee what thou art."

EATING AND DRINKING.

CHAPTER I.

THE OBJECT OF EATING AND DRINKING.

The object of eating and drinking is to impart heat and force to the system, and to supply its waste. All activity of the body—of the nervous as well as of the muscular system—is attended with waste. We cannot run, nor hop, nor skip, nor slide, nor bend, nor fall, nor breathe, nor hold our breath, nor lift a finger, nor hold a finger still, nor wink, nor weep, nor smile, nor attempt to repress our emotions; nor open or compress the lips, nor take a swallow of food or drink, without changing or consuming some portions of the body.

All mental activity is as much the result of bodily change as muscular. For the reasoning of the philosopher and for the love of the maiden; for the calm calculation of the scientist and the wild frenzy of the poet; for all fear, for all hope, for all fancy, for all adoration, there is a corresponding change and waste of the brain.

It is believed that for every time we break forth into passion, or breathe a prayer, there is waste of cerebral tissues without which such prayer or passion would have been impossible.

In the great economy of nature force answers to force and everything must be paid for.

Just as the candle grows shorter as it gives light; just as the coal on the grate is consumed while it warms the room ; just as in the galvanic battery the acid eats away the zinc in exact proportion to the quantity of electricity evolved, just so the brain wastes with every thought and motion.*

The writing of this work, of this paragraph, of this very sentence, costs more or less cerebral substance, which if not replaced will leave the author poorer than when he began it. The brain of the reader wastes as he turns over these pages ; if he accepts these views, waste will attend the emotion of acceptance ; if he rejects them, the feeling which induces such rejection will be attended by chemical changes in the brain, and additional argument will be arrayed against him. If he merely doubt these views, the act of doubting will cost him something of the most valuable part of his body, by a process that ought to aid in settling his doubts.

THE WASTE OF THE SYSTEM DURING LIFE NEVER CEASES.

Even in sleep these vast and complex changes do not cease. In the great human laboratory, the sound of the saw and the hammer is always heard by night as well

---

* The question of materialism we have here no space to discuss. The statements made above rightly represent the views of the most advanced cerebro-physiologists, and are admitted alike by those who favor and those who oppose materialism.

as by day, and only stops when its doors are closed for ever.   Whether the sleep be profound or broken the great tide of life rolls continuously, and every breath we draw is the signal for innumerable atoms to change their places.

That unknown and unknowable force which we call LIFE, is a great commander, and he controls a mighty and complex army.   When he wills, the nerves of vital force, travelling at the rate of 100 feet per second, carry his commands from the head-quarters in the brain to any one or to every organ; files of atoms range in platoons, form in line, right about face, deploy to the right and left, march slow or double quick, and an army which no man can number marches hither and thither, keeping time to the rhythm of the heart.

In short, the changes of the body are greatly diminished in activity, hence the necessity of sleep ; but they do not entirely cease.   The machinery is still going, but more slowly, and at less expenditure of force.

THE WASTE OF THE BODY MUST BE SUPPLIED.

Both the matter and the force of the body are limited.   Great is the difference between the weakest and the strongest specimens of humanity ; but the strongest are after all but weak, and both extremes and all the intermediate grades are alike dependent on the supply of matter and force that comes from food.

Just as one who continually draws on a limited account without making a deposit, must in time become poor ; just as the candle burns to the socket, and the lamp gives out because there is no more oil in it ; just as the fire on the hearth, unfed by fuel, grows fainter and fainter and goes out in darkness ; just as in the gul-

vanic battery the zinc must in time become so far consumed and corroded that it can no longer generate
an electric current, just so the body, if its ceaseless
wasting be not supplied, must grow weaker and smaller,
and in a few days must die.

It has been estimated that the body can be reduced
two fifths of its weight before starvation ; that up to
that point the body can feed on itself, on the reserve of
nutriment stored up in its own receptacles. As an average estimate, this is probably not far from correct ;
but undoubtedly the starvation point must widely vary
with the race, the climate, the state of health, the temperament and the power of will.

MATERIAL NECESSARY FOR THE GROWTH OF THE BODY.

From birth up to the age of twenty-five or thirty, or
even later, the body with greater or less rapidity and
constancy increases in size and weight. Between thirty
and fifty, and even later, even to extreme old age, there
may be increase of adipose tissue, though the stature is
usually fully attained before that period. There is little
doubt, also, that the brain may increase in size after as
well as fore thirty, and that this increase may go on,
at least till middle life is reached. Now in order to
carry on this growth there is as much need of material
as there is to evolve the forces of thought and motion ;
and if the material be wanting, or if it be meagre and
insufficient, the growth must either cease or be retarded.
Just as a plant deprived of its needful soil, moisture
and sunlight, becomes stunted and dwarfed, so the
body, deprived of necessary food, becomes weak and
sickly.

MATERIAL NECESSARY FOR WARMTH.

The heat of the body is maintained by the chemical changes that are constantly going on in it, and which are necessary for its existence and growth. Animal heat is a resultant of all or of many of these various changes and transformations, that never cease in the body so long as life remains. If, therefore, material must be supplied to replace what is lost in substance and in force, it must also be supplied to keep up the warmth that is generated by the active metamorphosis of tissues.

Without sustenance the body becomes cold, and the greater the amount of cold that the body suffers, the greater the amount of sustenance that it requires.

HOW THE FORCES, THE GROWTH, AND THE HEAT OF THE BODY ARE MAINTAINED.

The forces, the growth, and heat of the body are principally maintained by eating, drinking, and breathing. The air we breathe supplies the system with oxygen ; the water we drink supplies oxygen, hydrogen, and various mineral salts ; the food we eat supplies nearly all of the fourteen elementary substances of which the body is composed. Ordinary food also contains more or less water, is indeed largely composed of water ; and of many alimentary substances, it is the principal ingredient.

DIFFICULTY OF DEFINING FOOD.

The terms·that are most used, most familiar and for all practical purposes best understood, are oftentimes the most difficult to define. Food is a word over which there has been not a little dispute, and concerning its true and exhaustive definition there are broad differences of opinion. In the widest sense, it may include all, however taken into the system, that supplies its waste, maintains warmth and growth, and which corrects, intensifies, or economizes the vital forces.

Thus defined, it includes the air that we breathe, the solids that we eat, the liquids that we drink or absorb, as well as the active principles of most of the stimulants and narcotics, which are NEGATIVE food.*

Strictly speaking, all medicine is food, and all medical treatment, even external applications, act as food, either in a positive or negative sense.

We might go even further, and include under the term all those nameless hygienic and mental influences that serve to supply the waste of the body or intensify or economize its forces. Certainly it is in no way fanciful to call sunlight food, for it is indispensable to health ; and good news from a far country is often as truly food as water to the thirsty or as meat to the hungry.

The difficulty of accurately defining food comes from our ignorance of physiological chemistry. We know so little of the hidden processes of this body of ours, that we cannot well define what does and what does not minister to its substance.

For all practical purposes food may be used in its popular sense to include *those substances, whether liquid or solid, which, when passed into the system through the organs of digestion, sustain life.*

Food, as thus defined, I have divided into two classes— *Positive and Negative Food.*

Positive food includes the ordinary articles of diet,

---

* See my work on *Stimulants and Narcotics,* of this series.

both fluid and solid, that supply positive nutritive material to the system.

Negative food includes those substances that are embraced under stimulants and narcotics.

The subject of stimulants and narcotics is one of such great difficulty and importance that it has been thought worthy of separate and special consideration.

HUNGER—THIRST—STARVATION.

Hunger is the prayer of the body for nourishment. Hunger may be appeased either by solid or liquid food; thirst is satisfied only by water, either alone or in combination with solids or other liquids.

This prayer of the body for nourishment should not ordinarily go long unanswered. If either positive or negative food is not speedily forthcoming after the desire begins to be intensely felt, the system suffers.

One of the most terrible of deaths is that of starvation.

But although the sufferings of the starving are very great, it is not so much hunger as debility and pain of various kinds that causes the suffering. I was told by a medical friend, who for two days was lost in the woods without anything to eat, that the sensation of hunger soon passed away, and an attack of nausea and vomiting, resembling sea-sickness, came on that compelled him to rest for several hours. This attack was followed by painful debility but not by hunger.

Starvation from want of water, must be a death even more terrible than that from want of solid food, for thirst is a tyrant vastly more imperious and exacting than hunger. A fearful picture is drawn by Vambéry

of the agonies he bore in the Persian desert, from want
of water. A few days at most is as long as we can
exist without some form of liquid or solid, positive or
negative food.*

The stories that periodically arise of young girls who
live for weeks and months without food, may probably
be explained partly by fraud, and partly by ignorance.
At all events they do not, thus far, seem to bear rigid
scrutiny.

---

* In my work on *Stimulants and Narcotics*, I have given a number of remarkable illustrations of the sustaining power of alcoholic liquors, coca,
coffee, etc.

CHAPTER II.

WHETHER we accept or reject the Darwinian theory of development, we must, at least, concede that the present civilized races of the world have arisen by slow advances from the condition of semi or absolute savagery. In order to answer the question how man has found out what to eat and drink, we may well go back to the early struggles for existence, before civilization, as we now understand the term, had ever been begun. We may suppose, for example, that a portion of a tribe of about the condition of the native Australians, had been miraculously sustained by manna up to early manhood, and had been then let loose, we may say, on the African continent. In a few hours they would be hungry, and would seek for something to eat. They would be thirsty and would seek for something to drink. How would they find out what would be good and what bad, what palatable, and what nauseous? There would be no physiological chemistry to tell them, no one to analyze the berries, herbs, roots, barks, leaves, around them, and determine which were best designed for the human system.

Obviously there would be but one way of answering the question : *that is, by trial of whatever they could find.*

In settling the question of the healthfulness of any substance, the taste would be usually a reliable guide, since few natural substances that are grateful to the palate are injurious to the system.   But in order to learn the taste of any substance they must first try it, for neither the sense of vision nor the sense of smelling could be depended on.   The taste even could not be relied on entirely, for many substances that are more or less disagreeable, may yet be valuable as food, and may, in time, become delightful.

The same difficulties would be encountered in the selection of animal food.   Not until they had eaten the animal, or a part of it, cooked or uncooked, could they know what was good and what was evil.

For drink they would naturally experiment first of all with the water of the springs, lakes, and the ocean, and not until after trial would they know the difference between salt water and fresh, or hard water and soft. They would be just as likely to lie down on the beach and lap the waves of the sea, as by the brink of a moss fringed stream.   Their next experiment would likely be with the milk of animals or of trees, and with the juicy fruits.

As with vegetables and drinks, so with fruits ; the good would be discerned from the bad, only by trial, for neither the eye nor the smell would properly distinguish.

The observations of each individual would be communicated to the other members of the tribe, and their accumulated dietetic knowledge would be handed down from generation to generation.

As a result of this method of experimenting, it must necessarily follow—

1. That many must suffer in health, and lose their lives, by taking improper nourishment.

It is probable that many thousands have thus died that we might live.

When we consider that, even in our time, the accumulated wisdom of the ages does not prevent the occasional poisoning of a family by mushrooms, or by ill selected oysters, or lobsters, eaten out of season, or not properly cooked ; that meat, by parasites or otherwise, annually slays its victims ; that unripe and harmful fruits every season increase the rate of mortality ; and that many of these pernicious substances are eaten entirely without suspicion of their pernicious character, from their taste, or appearance, or flavor, we see that the reasons are strong enough for believing that our distant ancestors must have suffered more than can ever be estimated in their long struggle for existence, from want of dietetic knowledge. At first every mouthful was an experiment, and an experiment that might end in disease or death.

2. It would follow that the diet of the world would be modified by race, climate, soil, and the state of savagery or civilization. The past and present history of the world shows that such is the case.

The dietetic customs of the world are as opposite as their religious, or political, or social customs ; and among all nations of a progressive character there has been change in dietetic habit, with the change in civilization. "Half the world," it is said, "know not how the other half lives." The diverse customs of the world in eating and drinking will be described in a subsequent chapter.

As the object of food is to supply the waste of the body, it is obvious that the constituents of food must, to a certain extent, correspond to the constituents of the body, although the body is a laboratory whose subtle chemistry may and does variously transform the dissimilar substances that it receives.

ELEMENTARY CONSTITUENTS OF THE HUMAN BODY.

The number of elements of which the world is composed is *sixty-two*.

The number of elements in the human body is fourteen, as follows :

Oxygen,	Calcium,	Iron,
Hydrogen,	Sulphur,	Potassium,
Nitrogen,	Fluorine,	Magnesium,
Carbon,	Sodium,	Silicon.
Phosphorus,	Chlorine,	

The urgency, and frequency with which any element or proximate principle is required by the body, is to a certain extent proportioned to its relative importance. Oxygen, for example, is found in the body in great abundance, for it is one of the elements of water ; and we cannot live without breathing.

Thirst is more imperative than hunger ; we drink far oftener than we eat, and water is the principal constituent in all our food.

Phosphorus is one of. the most important and variously combined of all the elements ; hence its eminent importance in food and in medicine.

The elements of the human body are combined into the following

PROXIMATE CONSTITUENTS.

Water, which constitutes about two thirds of the weight of the body,
Albumen,
Fibrin,
Fat,
Gelatine,
Fluoride of Calcium,
Phosphate of Lime,
" " Soda,

Phosphate of Potash,
 " " Magnesia,
Carbonate of Lime,
 " " Soda,
Chloride of Sodium (common salt),
Sulphate of Potash,
Protoxide of Iron,
Sulphate of Soda,
Silica.

These elements and proximate principles must be supplied to the system directly or indirectly by the air we breathe, and the positive or negative nutriment that we take.

DIVISIONS AND PROXIMATE CONSTITUENTS OF FOOD.

The common varieties of food may be arranged under the following general divisions :

I.—CARBO-HYDRATES.

Alimentary substances that are composed of carbon united with hydrogen and oxygen, are called *carbo-hydrates.* The proportion of hydrogen and oxygen is the same as that in which they exist in water.

Under this head are included the following :

1. *Starch.* This is the most abundant material for nutrition that is known. It consists of grains that vary in size from $\frac{1}{700}$ to $\frac{1}{3500}$ of an inch in diameter.

Starch constitutes about 5 per cent. of turnips, 15

per cent. of the potato ; 60 per cent. of wheat, 82 per cent. of rice. It is found also in arrow-root and tapioca.

2. *Sugars*. Of sugars there are nearly a dozen different varieties. Among them are *Saccharose* or *cane*, found in sugar cane, beets, maple, carrots, pumpkins, etc.; *Glucose* or *grape* sugar, found in honey, figs, grapes and other fruits, and *Lævulose* or *fruit*, found in certain fruits, honey, and molasses. This latter does not crystallize.

Pectose. The pectose bodies are found in vegetables, in fruit, in roots, and in foliage. It is called vegetable jelly, and somewhat resembles starch and sugar.

Inulin. This substance is much like starch, and is found in several vegetables, as the dandelion and chicory.

Dextrin. This is found in very small quantities in several plants. In the process of malting corn it is produced from the starch.

Vegetable Acids. These are composed of carbon, hydrogen, and oxygen. Those which are principally known are *acetic acid*, produced from fermentation of the juice of plants ; *malic acid*, found in apples, plums, cherries, and other fruits, and in rhubarb ; *tartaric acid*, found in the grape, and pine and apple, etc. ; *citric acid*, found in the lemon ; and *oxalic acid*, found in the rhubarb, sorrel, and other plants.

II.—OILS AND FATS.

Carbon, hydrogen, and oxygen, make up oils and fats, but in a different proportion from the carbo-hydrates. The principal fats are—

> *Stearin*, found in tallow.
> *Olein*, found in oils.
> *Palmitin*, found in butter, wax, and oils.

All these consist of fatty acids combined with glycerine.

Of the elements that are necessary for nutrition, the fats and carbo-hydrates supply only three ; it is supposed that they cannot make flesh.

III.—ALBUMINOIDS.

The leading albuminoids are—

Albumen, found in plants and animal substances.
Fibrin, found in vegetables and animal substances.
Casein, found in the pea, the bean, and in animal substances.

The gluten of wheat flour is composed both of fibrin and albumen.

There are also other albuminoids that closely resemble each other.

The albuminoids contain nitrogen, sulphur, and phosphorus.   They are called *flesh-formers*

IV.—SALTS.

Under this division are included

Potash,   Soda,   Lime,   Magnesium,   Iron, etc.

These saline matters are found associated with the albuminoids.

Common salt comes under this division.   Water also contains salts of various kinds.

CHAPTER III.

In this chapter I purpose to give a very general description and history of the varieties of food that are most familiar to my readers, making no endeavor to be exhaustive.

ANIMAL FOOD.

Animal food serves an important and indispensable part in human nutrition. Life can be maintained without it, especially in temperate climates; but the highest activity of the brain and muscle is only possible for those whose diet consists in part of animal food. That man was designed to eat animal food, is shown :

1. By experience, and
2. Is confirmed by a study of his organs of nutrition.

The size, shape, position of the teeth, and the length of the alimentary canal, show as clearly as any fact in comparative anatomy can be made, that man does not deceive himself when he concludes from his experience that he is better, and stronger on a mixed than on a purely vegetable diet.

BEEF.—Of the different kinds of meat in common use, beef unquestionably stands at the head. It is the king of the meats. It is about three fourths water and one fourth solid matter. Fat is more nutritious than lean meat, but is more difficult of digestion, and there are very few in our climate who can take it in large quantities. The solid portion of beef is composed of albumen, fat, creatin, creatinin, inosinic acid, muscular tissue, and various salts.

MUTTON differs but little from beef in its composition, and is next the throne. The English mutton is far superior to the American, and in England mutton is more used than with us. The peculiar taste of mutton makes it less popular than beef.

BACON AND PORK.—The bacon of England and Ireland is superior to that of America, but it is nowhere the best kind of meat. It contains so much fat that only the hardy can digest it in large quantities. In the cities of this country it is but little used by those who are able to get anything better.

Pork, fresh or salted, is an article of diet that ought to disappear, and is disappearing before civilization. For centuries it constituted the leading article in the dietary of Europe, but with the advance of culture and the improvement in the race, the desire for it, and the ability to digest it, are both diminished. So thoroughly unfashionable is even fresh pork in our large cities, that a hotel or boarding house that should provide dinner in which pork should be the only meat, would soon be empty ; and fried salted pork is reserved for the extremely poor. In country places, among the farming

population, where the desire for this kind of meat and
the ability to digest it have not yet disappeared, and
where other meats are scarce or inaccessible, fried and
boiled salt pork is still the great dependence.

POULTRY, RABBITS, ETC.—The flesh of poultry, rabbits,
hares, is less nutritious than beef or mutton, but affords
a pleasing variety.

Wild fowl and game of all kinds are inferior to beef
and mutton, and some varieties are difficult of digestion.

VEAL AND LAMB.—Veal speaks for itself to those who
have been forced to subsist on it. Why it is that young
beef, which is so tender, is yet so often the cause of digestive disturbance, has never been explained.

Lamb is much more healthful, and is more palatable.

HORSE FLESH.—The use of horse flesh as an article of
diet is by no means a new experiment. It is said that
during the French Revolution the populace were fed for
six months on the flesh of horses, and no injuries re-
sulted, although severe complaints were raised against
it. Baron Larrey states that during his compaigns, he
gave horse flesh not only to the soldiers, but to the sick
and convalescent in the hospitals.

In 1835, in Paris, a commission of distinguished men
was appointed to determine whether horse flesh was
good food. They decided that it was palatable, and
that they "could not detect any sensible difference
between it and beef."

In 1841 horse flesh was openly sold at Würtemberg,
and since that time it has increased in popularity and
is sold both in Paris and in Germany.

Since that time, also, very many experiments have been made by the giving of banquets to settle the question of the palatability and the digestibility of horse flesh. The pretty unanimous result of the experiments is that it is sometimes quite difficult to distinguish roast horse from roast beef; that the horse soup was preferable to beef soup.

More recently still it has been proposed to popularize horse flesh as a measure of economy, and in order to help out the bad nutrition of the poorer classes, economy being regarded as its chief recommendation.

BONES contain considerable fat and nitrogenous matter. It has been estimated that 6 pounds of bones boiled for one hour, are equal to one pound of meat in nitrogen, and to nearly two pounds in carbon.

TRIPE contains 13 per cent. of albumen, and 16 per cent. of fat. It is easy of digestion, but a poor dependence for a meal.

BLOOD.—Pig blood is sometimes mixed with groats and fat, and the mixture is called *black pudding*.

EGGS.—There is in eggs about 25 per cent. of solid matter; of this 14 per cent. is nitrogenous, and 10½ per cent. is carbonaceous.

The yolk has 31 per cent. of fat. The white is mostly albumen.

Eggs are deficient in carbon. Here they are combined with bacon.

MILK.—Milk is the food that is designed for the young

of mammals ; but its value for adults has been greatly overrated.   Milk is composed of water, butter, sugar, casein, and mineral substances.   The relative proportions of the different substances vary in the different kinds of milk, but in no kind does the solid matter exceed 182 parts out of a thousand.

In the milk of woman the proportion of solids is 110.92 parts out of a thousand.

CREAM, which appears in the form of globules that gradually rise to the surface, contains about 35 per cent. of solid matter, and about two thirds butter.

WHEY contains the sugar and saline matter.   In *buttermilk* all the fat is excluded.

BUTTER consists of fatty and oily substances, saline matter, water, and casein.

The proportion of fat is from 70 to 95 per cent.

CHEESE is the *casein* of milk coagulated by rennet or by some acid.

The sugar of milk does not ferment like other sugars, else milk would be unfit for infants.   The milk of the ass, and the cow, and the sow, contains more sugar than that of woman.   Goats' milk contains more butter but less sugar than that of woman.

THE COW TREE.

Corresponding to the bread fruit, there are in certain warm countries *milk producing plants* which are found in a certain district of South America.   The milky juices of plants are generally acrid, bitter and injurious as food ; but that of the cow tree, which is obtained by making incisions in the trunk, is described by Humboldt as "tolerably thick, devoid of all acridity, and of an agreeable balmy smell."

Humboldt further says that "amidst the great number of curious phenomena which I have observed in the course of my travels, I confess there are few that have made so powerful an impression on me as the aspect of the cow tree."

The same authority further states, that "this cow tree, like ordinary cows, is milked in the morning; for then the fluid is most abundant." Then "the negroes and natives are seen hastening from all quarters, furnished with large bowls to receive the milk, which grows yellow and thickens at its surface."

Caoutchouc, or rubber, is a kind of vegetable milk, since it contains caseous matter.

VEGETABLE CHEESE.

When the milk of the cow tree is exposed to the air, the yellow cream rises to the top, and a substance is formed resembling cheese, and is so called by the natives.

BUTTER TREE.

Mungo Park speaks of a tree in Bambarra that gave an oily substance resembling butter; and the same is mentioned by more recent travellers.

FISH.

Among the white varieties of fish are whiting, cod, haddock, sole, flounder. These contain only about 22 per cent. of solid matter, and of this 18 per cent. is nitrogenous.

MACKEREL, EELS, SALMON AND HERRING.—These contain much more fat.

All fish are best at the time when the roe ripens. At this time they are fatter, richer, and have a better flavor.

SHELL FISH.—Shell fish contain about 13 per cent. of solid matter.

Oysters are the easiest of digestion. Lobsters, crabs, scollops, mussels, etc., are more difficult to manage, especially when not well cooked, and the latter sometimes cause serious injury.

TURTLE.—The flesh of the turtle contains much less fat than is supposed. It is about three fourths water, and fat constitutes only about $\frac{1}{2}$ per cent. of its solid ingredients.

Genuine turtle soup is more digestible than the so-called mock turtle.

CEREALS.

WHEAT, RYE, BARLEY AND OATS.—The early history of these cereals is not fully known. It is stated that the inhabitants of the lake dwellings of Switzerland cultivated the cereal plants. But as Darwin remarks, "botanists are not universally agreed on the aboriginal parentage of any one cereal plant." It is stated that all our cereal plants are cultivated varieties of wild species that have been found in Asia.

There is little question that man very early began to cultivate the earth, and that from a very distant period these cereals were used by him as food.

Of all the cereals *wheat* is the most used.

Oatmeal does not make bread. Oatmeal porridge is, however, a leading article of food with the peasantry of Scotland, Ireland and England, and was formerly used by the wealthy and luxurious.

Barley flour, unless mingled with wheat flour, does not make good bread, and, therefore it is chiefly used in the form of porridge.

Rye flour is inferior to wheat flour, but it is much used in connection with it. Rye bread is common in Germany, and in the United States is used not a little, alone, or mingled with Indian meal or wheat flour.

BREAD.

Bread is made of several varieties of cereals; but wheat bread is the form that is most used.

The wheat is ground and sifted. Wheat contains from 60 to 90 parts of flour, to from 10 to 40 parts of bran. The average is about 10 per cent. of bran.

There are several grades of flour. The *first* is *fine flour*, and contains no bran; the *second*, called *canaille* or *seconds*, contains a little bran; the *third*, called *shorts*, contains a larger percentage of bran; the *fourth*, called *pollards*, is almost all bran, and is used mostly for animals.

Unbolted wheat flour makes, what is called in Europe, "brown bread," and in America "Graham bread," after Mr. Graham, the founder of "Grahamism." It is less digestible than bread made of fine flour, but it contains some important nutritious ingredients in the bran, and is more laxative than the bolted.

Bread contains about 60 per cent. of solid substances, 55 to 58 per cent. of dextrin, gum, starch and sugar, 1½ per cent. of fat, and 8 to 10 per cent. of nitrogenous substances, of which starch is the leading ingredient, and 40 per cent. of water.

The proportion of water varies with the age of the bread, and the method of making.

MAIZE OR INDIAN CORN.—Indian corn is without doubt of American origin. It was used by the aborigines in both North and South America, and two kinds of it have been found in the tombs, "apparently prior to the dynasty of the Incas."

Very numerous varieties have arisen. Indian meal is made into a bread that is called "brown bread," or "corn bread;" it is a favorite in New England and in the Southern United States. In New York and the Middle States it is but little used, and is there known as the "Boston brown bread." It is healthful, nutritious, contains a good proportion of fat. It is more laxative than wheaten bread, but is less adapted for feeble digestion.

BUCKWHEAT FLOUR in the form of "buckwheat cakes" is distinctively an American luxury. They do not contain all the elements of nutrition, but usually, when properly made, are easy of digestion, and as beneficial as they are delightful. Their deficiencies are generally made up by the molasses, or sugar, or butter, or cream, or milk, or gravy, or meats with which they are eaten.

RICE is in some features nutritious; yet it is not the best food, although millions subsist on it. It is deficient in important elements of nutrition. It is better adapted for warm than cold climates. In cold and temperate regions, and for hard workers with brain and muscle in any climate, it does not sufficiently exercise the stomach, nor warm the system, nor feed the brain or muscle.

VEGETABLES.

POTATOES.—The potato is an American plant. The cultivated varieties differ but little from the wild species. It is stated that in Great Britain 125 kinds of potato exist.

The potato contains about 25 per cent. of solid matters, four fifths of which are carbo-hydrates, and the remaining fifth fats, albuminoids and salts. The potato alone, therefore, can sustain life, but it is very deficient in fat and in the phosphates.

It is the king of the vegetables, but as an exclusive diet it is to be deprecated.

SWEET POTATOES.—This vegetable, which is much used in the South and the West Indies, is hardly as digestible for the majority as the so-called Irish potato. It is, however, very palatable, and need not be feared by those to whom it has not in some way shown itself inimical.

YAMS.—The yam, so much eaten in the West Indies, resembles the sweet potato. It is quite nutritive. When ground into flour it is made into bread. In the same way ordinary potatoes are in America sometimes mingled with Indian meal to make brown bread.

PLANTAINS.—The plantain, though a fruit, is to the tropics what the potato is to the temperate regions. In large districts of Africa, it is a great dependence, and wines or beers are also made from it.

TURNIPS, PARSNIPS AND CARROTS.—These vegetables

contain a larger percentage of water than the potato, and are of much less value.

They are quite palatable, and help out the variety in a dinner.

CABBAGE.—The cabbage is believed by many to be descended from a wild plant that grows on the western shores of Europe, while others contend that its ancestors are to be found on the Mediterranean.

The cabbage contains only about 5 per cent. of solid matter. Like all other vegetables its value in diet consists of its saline ingredients.

PEAS AND BEANS.—A kind of pea now extinct has been found in the lake dwellings of Switzerland.

The common garden pea is believed to be descended from a variety that has been found in the Crimea.

Peas and beans are valuable for their legumin, which resembles casein of milk, and which constitutes nearly one fourth of their weight.

Their nutritive value has been much overrated.

BEETS.—Beets contain a considerable percentage of sugar, and are, therefore, nutritive, but they are only of value in connection with other food.

ONIONS.—Onions contain nourishment, and in the opinion of some are superior to any other vegetable. Raw onions are usually disagreeable to a delicate stomach.

Leek, garlic and *shallot* are very similar to the onion. All of these vegetables are more used by sailors than by landsmen, and in the West Indies, and in warm climes generally, than in the North.

CUCUMBERS.—Cucumbers, whether eaten fresh or in the form of pickles, are of but little value, except to give variety and to stimulate the appetite. To some temperaments they are indigestible and must be used sparingly or wholly abstained from. Those who taste them over and over again after they have eaten, may as well do without them—and they can do so with the consolation that they are losing but little.

PUMPKINS AND SQUASHES.—These when made into stews contain a small amount of nutriment, are quite palatable, and are rarely otherwise than a beneficial accompaniment to more substantial food.

TOMATOES.—Tomatoes were formerly regarded as only fit for hogs, and their introduction to fashionable tables in this country in raw and stewed form is of very recent date.

Their value as medicinal food has probably been much overestimated, and is based chiefly, I think, on the same theory that has given Graham bread such great prominence, namely, that the value of any article of food is in pretty exact proportion to its disagreeableness, since, to the majority of persons, tomatoes are at first repulsive.

There is no evidence that they are in any respect superior to ripe peaches, or pears, or apples.

Tomatoes are said to contain considerable oxalic acid.

FRUITS.

THE PEACH.—The peach is a native of Persia. Originally it was a kind of bitter almond, but by long culture has developed into the luscious fruit as it is now known. It was carried into Italy by the Romans, and thence into France. It was introduced into England about the middle of the 16th century. At first it was thought to be injurious on account of the prussic acid contained in it.

There are 20 known varieties of the peach, and 20 varieties of the nectarine, which closely resembles it.

The peach is one of the most healthful as well as the most agreeable of fruits. It can be eaten in enormous quantities without injury, and oftentimes with positive benefit.

APRICOTS.—The tree on which apricots grow is descended from a species that is found wild in the Caucasian region. It is estimated that there are 12 varieties.

PLUMS.—The plum tree is supposed to be descended from the *bullace*, which is found wild in the Caucasian region, and .in Northwestern India. There are about 40 varieties.

CHERRIES.—Cherries are believed to be descended from several wild stocks.

APPLES.—Whether the apple is descended from one or from several closely allied wild forms, is a matter of doubt. There are now nearly 1,000 varieties.

PEARS.—Pears are descended from one form that is found wild. Both pears and apples, when ripe, can be liberally used without fear by the majority of those in average health. There is need, however, of greater caution than in the use of peaches.

STRAWBERRIES.—In 1246 only three varieties of the strawberry were known. "At the present day the varieties of the several species are almost innumerable," and they have greatly increased during the past half century. With culture it has wonderfully increased in size and quality.

GOOSEBERRIES.—The gooseberry plant comes from a wild form that is found in Central and Northern Europe. There are now about 300 varieties. Under cultivation it has greatly increased in weight. The present weight has been known to be seven or eight times the weight of the wild gooseberry.

WATERMELONS, MUSKMELONS AND CANTELOPES.—Watermelons are rightly named ; they contain little besides water. In that little is oftentimes enough to induce serious illness.

Like cucumbers, they are treacherous fruits, and sometimes betray their best friends.

Muskmelons and cantelopes are more trustworthy and are more valuable.

All these fruits are less liable to do injury when eaten as a part of the regular meal.

CRANBERRIES.—A very pleasant and healthful acid is found in the cranberry. When stewed with sugar they are generally harmless, and by those of delicate digestion are sometimes eagerly craved.

WHORTLEBERRIES, BLACKBERRIES, ETC.—These berries, like peaches, can usually be taken in large quantities with only agreeable results.

The medicinal character of blackberries, like that of
tomatoes, has, I suspect, been somewhat exaggerated,
but that they are of great value in summer food there is
no question.

FRUITS NOT VERY NUTRITIVE.

The amount of actual nutriment in many of the most
luscious fruits and vegetables is exceedingly small, and
enormous quantities must be eaten in order to make a
meal of them, and of some varieties no amount that can
be swallowed will very long satisfy the appetite.

To judge of the value of food by the *amount of its
nutriment* merely, is as unscientific and as unphilosoph-
ical as it would be to judge of the character of men by
their weight avoirdupois.

These watery and luscious fruits and vegetables serve
not only to quench the thirst, to supply acids and sugars,
etc., to the body, but also to stimulate the appetite for
other and more substantial food, and to assist its digestion, and very likely to enhance, in ways that are past
our finding out, its assimilative power.

Fruits do not contain more than five per cent. of solid
matter, and very little of the albuminoids ; but they
contain considerable sugar and acid. They are valuable
in diet for their saline ingredients, for their sugars and
acids, and for the flavor and zest that they give to other
food.

Ripe, luscious fruits of nearly all the leading varieties, can be eaten in their season, by those in average
health, in *enormous quantities*, to the full extent of the
appetite and between meals, not only without injury
but with very great benefit. To this rule there are
individual exceptions.

TROPICAL FRUITS.

The value of tropical fruits, like the beauty of tropical flowers, has in some respects been greatly exaggerated.

Some of the flowers that grow in southern lands are indeed beautiful, and a few of them are magnificent, but the average effect of a tropical forest or landscape is far less attractive than that of the gardens, the fields, and the hedges of the North. Just so, some of the fruits of the tropics are rich and tempting, and one or two varieties are luscious beyond anything that we have in our colder clime, yet as a whole they bear no comparison with the produce of the orchards of England or America. Between the warm and the cold climes there is in this respect less difference than has been supposed. The grape, the peach, and the apple, will outweigh in the variety of uses nearly all the fruits of the tropics.

The best of the tropical fruits are the *orange*, the *durion*, the *litchi*, the *mangosteen*.

ORANGES.—The orange is to us as familiar as the apple, and strange as it may appear, costs less than the apple in the New York market.

In the South, oranges, and indeed other fruits, are eaten early in the morning, soon after rising, or as a part of breakfast. The old Spanish proverb, "Fruit is gold in the morning, silver at noon, and lead at night," could only have originated in a warm clime, for it certainly does not apply to the latitude of Northern Europe and America.

Except in the warmest weather, fruits are not desired at breakfast, and, to borrow a medical phrase, are not indicated.

In the Southern States oranges and melons are morning food ; and in Texas I observed that cantalopes were, in their season, on the breakfast table almost as regularly as the bread or coffee.

THE DURION.—Of this fruit, we have a most enthusiastic description by Mr. Alfred Wallace. He states that in the Malay Archipelago it is regarded by the natives as superior to all the many varieties with which they are favored.

It grows on a lofty tree, somewhat resembling an elm. The fruit is about the size of a large cocoanut, is of a green color, and is covered with short spines, and is so heavy that when it falls from a height on any individual it produces a terrible wound.

"The pulp is the eatable part ; its consistence and flavor are indescribable. A rich butter-like custard highly flavored with almond, gives the best general idea of it, but intermingled with it come wafts of flavor that call to mind cream-cheese, onion sauce, brown sherry, and other incongruities. It is neither acid, nor sweet, nor juicy, yet one feels the want of none of these qualities, for it is perfect as it is. It produces no nausea or other bad effect, and the more you eat of it, the less you feel inclined to stop. In fact, to eat durions, is a rare sensation worth a voyage to the East to experience."

Rev. Mr. Mellen, formerly U. S. Consul at Mauritius, informs me that an exceedingly luscious fruit also grows there, which receives the name of "bull's heart." From his description I should judge that it was much like the durion. He says that it is as much better than the finest custard as custard is better than hominy.

Mr. Wallace further says, "It would not perhaps be

correct to say that the durion is the best of all fruits, because it cannot supply the place of the sub-acid juicy kinds, such as the orange, grape, mango, and mangosteen, whose refreshing and cooling qualities are so wholesome and grateful; but as producing a food of the most exquisite flavor it is unsurpassed. If I had to fix on two only as representing the perfection of the two classes, I should certainly choose the durion and the orange as the king and queen of fruits."

While our mouths water over the descriptions of fruits that can be tasted only by those who make long and wearisome voyages, we can find consolation in the possession of the peach, the pear, and the strawberry.

BREAD FRUIT.—The bread fruit tree, which grows, though never abundantly, in many tropical regions, is one of the most remarkable objects in nature. The fruit is of the size of a melon and is of the consistence of pudding. It may be baked, stewed or fried. It is eaten with gravy, with sugar, molasses, butter, or milk. Its taste is compared to that of mashed potato and milk, or when cooked with sweets, to a delicious pudding. Like bread and potatoes it becomes a standard food which is always useful, never injures, and of which we never tire.

LEMONS AND LIMES.—The acid in lemons and limes is not only useful against scurvy, but fulfills an excellent purpose in combination with fish of all kinds, and in the form of lemonade.

Lemonade is a drink that for the majority is as healthful as it is agreeable.

PINEAPPLE.—This very palatable fruit is less digestible than the orange, and must be eaten with caution, especially in debilitating weather.

BANANAS.—The banana contains some little nutriment, and is not often injurious.

COCOANUT.—The cocoanut, when it arrives in this country, is so hard that it becomes indigestible, and the milk is not particularly refreshing. It has been said that in those regions where they grow, the milk is much better and the nut itself particularly delightful. This statement does not correspond with my experience. I have taken cocoanuts from the tree and eaten them on the spot, and failed to discover the superiority that has been claimed for them when perfectly fresh, and the milk when drunk directly from the nut was to me rather sickening, although others were more pleased with it.

For myself I should prefer the cocoanut after it has been kept for a number of days.

I may say also that the flavor of the orange, when eaten fresh from the tree, has been similarly exaggerated by travelers. Oranges, when taken ripe from the tree, remain luscious for a considerable time.

FIGS, RAISINS, DATES, PRUNES.—All these dried fruits are useful as well as palatable in their way, that is, as relishes or in combination with other food. They are injurious only to weak stomachs. The habit of forcing them down in large quantities in order to keep the bowels free, is abominable. When eaten in large quantities it is well that they should be combined with some acid fruit, or at least with bread or meat.

The idea that figs are better when fresh from the tree on which they grow, is, if I may generalize from my own experience, another delusion. I have often eaten them when recently plucked and prepared for the table, and found them simply tolerable.

It has become so fashionable for superficial observers to declare that the unfortunate dwellers in the temperate zones, know nothing of the lusciousness of the tropical fruits, that it is well to stop and inquire how much of truth there may be in the statement.

NUTS.

The walnut tree grows wild in the Caucasus and Himalaya. There are a number of varieties. Nuts of all kinds contain considerable fat, and are hard of digestion. They are best taken in small quantities, and in connection with other food of a different character, such as fruits. They do less harm *after meals* than on an empty stomach or just before retiring.

CHESTNUTS.—Chestnuts contain a considerable nutriment, and by the peasants of the Apennines are sometimes depended on for food. They are less burdensome to the stomach than walnuts.

SUGAR.

Sugar and molasses produce heat and fat, and also sustain the system in other ways that are not understood.

They are much longed for by children, and they do

them good. They are palatable—the strongest possible argument in their favor—and they supply the fat and the warmth that children need.

The popular candies are frequently injurious, on account of the bad character of the substances that are used to color them.

ICE CREAM.

Ice cream, when made of pure milk or cream, and flavored without the aid of injurious ingredients, is not only harmless, but positively beneficial.

It is very properly eaten in the evening or between meals, or at least as an accompaniment of hot food.

An objection to eating it largely at meals, is that it cools the stomach, especially if it be delicate to such a degree that digestion is hindered.

The custom of combining ice cream with acid fruits is as wise as it is agreeable.

WATER.

Water constitutes, as we have seen, the principal portion of our food. It is the leading constituent of all our so-called solid food; it is used by itself and in all our beverages.

Pure water is never found, and only with great difficulty is it manufactured approximately pure by the chemist. It is without odor or color, and is therefore not adapted for drinking purposes.

The purest of all water is rain water; but this always contains more or less of organic matter, with nitric acid, ammonia and salts.

The ingredients of the majority of our drinking waters, are oxygen, nitrogen, carbonic acid.

The waters of *mineral springs* contain, besides these substances, various salts of potash, iron, and soda, on which their medicinal virtues depend.

Some waters contain as much as 15 or 20 cubic inches per gallon of carbonic acid gas, and water thus impregnated are much more palatable than those that are entirely free from this gas.

Hard water is that which contains salts of lime, potassium, sodium, or magnesium. These salts are found in some waters in the proportion of from 10 to 15 grains to the gallon.

It is believed that a moderate proportion of salts in our drinking water is beneficial, although hard water is very inconvenient for culinary or washing purposes, because the fat of the soap combines with the salty matters in the water, and forms insoluble soap.

Distilled water, though almost perfectly pure, is not agreeable to the taste, until it has been thoroughly agitated and mixed with common air and carbonic acid gas. This may be done by pouring it from one vessel to another many times, or by forcing common air into it.

The water of rivers frequently contains a considerable percentage of organic matter. The water of the Mississippi, for example, when poured into a tumbler, is exceedingly turbid, and if allowed to stand for a few moments, a considerable quantity of earthy matter settles on the bottom.

It is strange how freely this water can be used in many instances with apparent impunity. The boatmen on the Mississippi drink as much as they want of it without fear.   That it sometimes causes affections of the bowels there is no doubt ; and yet I have had opportunity to see that bad results occur only in exceptional cases.

Water should be drunk freely during meals, and as often as thirst requires.

The doctrine that we shall not drink during meals because some animals do not, is so abominable, so unscientific, and so cruel, that it is worthy of being classed with the dogmas of the medieval ascetics.

SODA WATER.

Soda water is justly regarded as a peculiarly innocent beverage.

Very few are harmed by it, and thousands in the country are refreshed and positively nourished by it.

This compliment applies only to good soda water.

The art of making soda water just what it should be is underrated, or, at least, is practised only by a few.

Those to whom soda water causes nausea or sick headache, need no advice of mine to abstain.

CONDIMENTS.

Condiments are substances that sharpen the appetite and stimulate the digestive organs, and aid in the digestion of ordinary food.   The best known condiments are :

Pepper,	Nutmeg,	Mace,
Cayenne Pepper,	Cloves,	Cassia,
Mustard,	Allspice,	Turmeric,
Cinnamon,	Ginger,	Curry Powder.

Most of these condiments are natives of warm regions ;

and it is a fact of experience that they are most needed, most desired, and most useful in those countries.

They are used habitually in cold and temperate climates, but not as a rule in large quantities.

THE PEPPERS which grow in the West and East Indies, consist of a volatile oil, a crystallizable substance called piperine, or capsicine, starchy matter, gum, woody fibre, and salts of potash.

GINGER is cultivated in tropical Asia and America. It consists of a volatile oil, resin, gum, starch, woody fibre, acetic acid, sulphur, salts of potash, lime, and iron.

TURMERIC.—This is the tuber of a plant that grows in India and China. It has a peculiar aromatic taste, and is of an orange yellow color. Turmeric is a leading ingredient in curry powder.

CURRY POWDER consists of a number of ingredients— among which are black and cayenne pepper, coriander seeds, ginger, allspice, cloves, turmeric, cardamom, fœnugreck and cumin. Of these turmeric, coriander seeds and black pepper, are the most important. Coriander seed and cardamom are sufficiently familiar. Cumin seeds strongly resemble caraway. Fœnugreck seeds are the produce of a plant that grows in India, Arabia and Sicily.

Condiments are used in hot climates to an extent that greatly surprises travelers. Pepper and curry in enormous quantities that a Northerner cannot endure, are there used in combination with more substantial articles. They seem to aid digestion, to have a cooling effect, and to compensate for deficiency in salt.

COMMON SALT.—Common salt is the most universal food of man.

Salt was used by the Jews in their sacrifices, and by the Arabians was placed on the table as a special mark of hospitality. The Hindoos swear by salt, and the Abyssinian gentleman carries a piece in his pocket, and takes it out and offers it to a friend to lick as a mark of respect and esteem.

Among the Apingi tribe of Africa salt is so scarce that ten pounds will buy a slave, and Du Chaillu thinks that the ulcers and diseases of the skin to which these people are subject are largely caused by deficiency in this article.

The deprivation of salt is one of the most terrible of calamities. In Africa tribes have been found who will sell a wife or a child for a small quantity.

Livingstone says that the poor Backwains who are forced to eat roots without salt are troubled with indigestion. Milk and meat are in Africa found to be good substitutes for salt. Of all the continents Africa seems to suffer most from salt famine.

It is not very complimentary to the state of the science of physiological chemistry to be obliged to confess that the function of an agent so indispensable as salt is not understood. When we have stated that it aids in the digestion of other food, and in the processes of absorption and secretion, we have said all that we know and perhaps more than we know.

SUBSTITUTES FOR SALT.

It seems that these elementary substances, such as meat, milk, etc., which contain salts of potash and other mineral matter, are, to a certain extent, substitutes for salt ; that when these substances are used in abundance less salt is needed, and the want of it is better endured. Pepper also is said to have the same effect. The appetite for salt, as we all know, depends on the quality of our food.

CHIVI.—The Indians of Javita, South America, prepare a substitute for salt, of muriate of potash, caustic lime, and the earthy salts.

CUPANA, CHINIA, CIRO.—The Indians of Javita also make a salt by reducing to ashes the seeds of the plant cupana. The seeds also are carefully scraped, mixed with cassava, covered with plantain leaves, and allowed to ferment.

Chinia, another salt, is made by reducing to ashes the spadix and fruit of the chinia palm tree.

In Madagascar salt is obtained from the sap of the ciro palm tree.

The milk of the cocoanut tree also contains saline matter.

We who live in a land where salt is not only cheap, but the very cheapest article of food, so that not to be able to earn it is regarded the expression of extreme poverty, can with difficulty realize the condition of the inhabitants of Central Africa, where only the very richest can afford to have salt at their meals ; where children, when they can get it, suck it like sticks of candy.

RELATION OF FOOD TO WORK.

Various and interesting experiments have been made to determine the relation of food to the amount of work that it is capable of accomplishing. Analysis of the different alimentary substances, will tell us the percentage of carbo-hydrates, of albuminoids, and so forth, but it does not tell us just what percentage of these are assimilated, nor can it explain the manifold changes which these must undergo in the body. Hence any estimate of the amount of work that food is capable of, based simply on their chemical constitution, would be imperfect and erroneous. The usual method of determining the relation of the different kinds of food to work is to estimate the amount of *heat* they give off when burned.

Heat is regarded as an equivalent of motive force, and the food that gives off the most heat is the best for giving heat and motive power to the body. The estimates thus found must, at best, be only approximate. They make no allowance for the changes that may take place in the body, and to which I have referred. Moreover, *cellulose* of vegetable food is indigestible, though it burns readily and evolves much heat. On the other hand, the albuminous elements of food do not give out their entire force in the body. Making allowance so far as possible for these facts, we may arrive at an estimate of the work power of food that may, perhaps, be approximately accurate.

Thus it has been estimated that the latent force of 700 grains of food—less than 1½ ounces of butter—is sufficient to raise 1,000,000 pounds one foot high in one day.

Practically it is found that only about half of the latent force of food is evolved in the body. In all mechanical arrangements—such as the steam engine—the proportion of wasted force is ten times greater than the force that is utilized.

It has been estimated that three fifths of the force evolved by food is used in the acts of respiration and the movements of the heart.

It is believed that the leading function of nitrogenous food—meat, fish, bread, etc., is to repair the tissues of the body. This view is rendered probable by the experiments on animals, which show that with a purely carbonaceous diet—fat, oils, sugar and starch—the weight is rapidly lost.

That it is possible to live, after a sort, on a diet very largely carbonaceous, is proved by the experience of the wretched millions of China and India.

Then again experience demonstrates that it is not enough that the food should contain nitrogen : it must contain it in a variety of combinations ; hence the necessity of a *variety* of food.

It is not improbable furthermore that nitrogenous substances may aid in the digestion of the carbohydrates.

It is certainly proved that articles of food, which are the same in chemical constitution, differ widely in their capacity to sustain life.

Carbo-hydrates—fat, starch and sugar—not only serve to warm the system, but they also repair the tissues, especially the adispose tissues. The heat producing power of fat is believed to be three times greater than that of starch and sugar.

They also assist in the digestion of nitrogenous food. Here, then, is another argument for *variety* in our diet.

The question whether the tissues of the muscles are actually consumed in muscular exercise or only the blood, has received various answers.

The belief has been that force is developed mainly

by the oxidation of the carbo-hydrates in the blood, and that when abundant food is given the muscular tissue is not changed.

Prof. Austin Flint, Jr., thinks he has shown, by very careful experiments made on Weston, the famous pedestrian, during one of his greatest feats, that very severe exertion is attended with " consumption of the muscular substance." Whether this consumption of the tissue of the muscles takes place during the ordinary muscular exercise, when the food is various and abundant, is yet to be shown.

Finally, I may remark that the belief is not unreasonable that the animal heat is produced by the transference of the potential energy of the food in the blood into muscular power.

CHAPTER IV.

The facts contained in this chapter have been obtained from a pretty exhaustive study of all possible sources of information concerning the eating customs in all parts of the world, where man has penetrated, and the generalizations are drawn from a careful survey of the facts considered in the light of recent science.

1. *The customs of different peoples in regard to eating and drinking are as diverse and opposite as their customs in regard to religion, government, and social life, and nearly all forms of organic life, when used as food, seem to be capable of sustaining life.*

I have only space, under this head, to refer to a few of the more noteworthy diversities of custom.

In Persia, horse flesh and camel's flesh are the great dishes.

The Tartars live on milk, butter, and oatmeal. The Thibetans, rich and poor, are meagerly fed. They have little rice, or wheat, or fruit, and depend on black barley; from which they make tsamba, their leading article of diet

The staff of life in East Africa, is *ugali*, a porridge made of boiled millet, or maize flour.

The North American Indians live on the flesh of animals obtained in hunting, fresh water fish, and others obtain corn and fruits.

The African Bushmen eat roots, bulbs, garlic, the larvæ of ants, locusts, grasshoppers, lizards, and anything else they can get hold of.

The Hottentots pick vermin from each other, and eat them as a matter of course, declaring that the best way to get revenge for their eating, is by our eating them.

The Australians eat the prairie rat, the flying fox, and the ant eater. The Malayans delight in a species of bat.

The Bengalese eat cow or buffalo beef, pork, pigeons, and in times of scarcity, snakes, frogs, ants and rats.

Enormous fish, that we know nothing of, are caught in tropical rivers. I can only mention the "Baggar," the "Bayard," and the "Coor;" the catching of which is so graphically described by Baker.

The Chinese eat sea slug, which is regarded as a great delicacy; and birds' nests, sharks' fins, and fish roes. The poorer classes subsist mainly on rice and fish; but sometimes resort to the heads and entrails of fowl, earth worms, mice, dogs, cats, frogs, horse-flesh, unhatched ducks, and snakes, and rotten eggs.

The popular conception that dogs, and rats, and mice are a prominent article of food in China is erroneous.

In this country who would think of eating water-melon seeds, except when on the borders of starvation? But Mr. Huc tells us that in China they are an object of daily consumption. They are taken before meals, and between meals; and when it is desired to treat a friend or caller. Everywhere they are sold ; and children run to invest their little money in them like so many sugar-plums. "The consumption of them throughout the empire is something incredible, something beyond the limits of the wildest imagination."

A merchant who was invited to a feast of the Greenlanders, counted the following dishes : "Dried herring; dried seal's flesh, called mikiak ; boiled auks, part of a whale's tail in a half putrid state, which was considered a principal dish ; dried salmon ; dried reindeer venison; preserves of cranberries mixed with the chyle from the maw of the reindeer ; and lastly, the same enriched with train oil."

A ROYAL FEAST IN AFRICA.

Du Chaillu gives an interesting description of a royal feast that was given to him by King Obindji, on the banks of the Ovenga River, in Equatorial Africa. To the furious beating of drums, twenty of the king's wives came forth, each bringing a dish which they placed on the mat. The first course was *boiled elephant,* that had been cooking since the day before, to make it tender. Next appeared *boiled crocodile,* the broth of which was well seasoned with lemon juice and pepper.

A *monkey* roasted whole followed, wonderfully resembling a roasted baby. A *wild boar* was the next surprise, with a strong flavor as if it had been killed many days before.

Then came the less astonishing *buffalo ribs* and *tongue,* which was followed by *smoked otter, antelope,* and the light *gazelle.*

Instead of bread, plantains were placed in a large dish in the center.

The meats were followed by the fish, as *manaters, mullets,* and *turtles.*

As a kind of second course, *boiled gorilla* was prepared by a friend of Du Chaillu, and "also a dish made of part of a large *snake* cooked in leaves."

Du Chaillu gives his impression of each article. The elephant was abominably tough ; the crocodile soup was not at all agreeable, although the taste was extinguished by pepper ; the wild boar was putrid, but the Africans "thought it exquisite ;" the otter he did not like, and the buffalo he hardly touched ; but the monkey was "perfectly delicious," as good as the best of venison.

The Arabs almost depend on "*waker,*" which is called in Ceylon and India "Barmian," and "Bandikai." The pods of this are dried, and are mixed with almost every kind of food. Besides this they have the "regle," a juicy, brittle plant.

A great staple with the Arabs is "dhurra," a kind of grain, of which they make "melach" and "abrey" cakes, which they eat on their long journeys.

In other lands all sorts of strange animals and birds are eaten—*parrots,* which when old are tough, but when young are like the pigeon ; the *wild boar ;* the *ass,* which has been highly valued ; the *donkey,* not unlike the horse ; the *hyena ;* the *leopard ;* the *reindeer ;* the *tiger,* which the Malay believes to be a specific against disease ; the *panther ;* the *wild cat ;* the *bear ;* the *lion* and the *jaguar ;* the *beaver ;* the *kangaroo,* the tail of which makes a soup in comparison with which that of ox tail is tame enough.

Nature is continually surprising us with singular freaks and caprices in her productions. In the interior of Brazil *sweet lemons* are found, which resemble in outward appearance, the ordinary sour lemon.

Sour honey is also produced by a certain species of bee in South America.

There are many tropical fruits that rarely, if ever, reach the North, but are known only to travelers. Among these I may mention *lansat, jambon, jack, rambutan, blimbing,* and *mangosteen* which is by some regarded as the best of the tropical fruits.

In the West Indies the *alligator pear* is much valued; Brazil produces the elegant *custard apple.* In the West Indies also grows the *mamma apple.* A fruit of Guinea is the *akee.* One of the finest luxuries of the East, is the *litchi,* which ranks nearly on a par with the durion and mangosteen.

Other fruits of warm climates are the *lucuma, maqui, hog plum, sapota, star apple, chupon, guava,* and *tamarind,* whose peculiar acid is so delightful to invalids.

Among productions of the tropics that are little known among us, may be mentioned the *koola nut,* which is said to be very nutritious; the *saloom* and *nebbuk,* that grows in Egypt; *motsouri* and *nijii* beans; the *mogametsa,* a bean with a little pulp around it; the *moribi,* a bright red bean, and the *manioc* and *plantain,* which are to Africa what bread and potatoes are to Europe and America.

Travelers speak also of the *marolo,* "small and wholesome, and full of seeds, like the custard apple;" the *mamosho,* or "mother of morning," described as "juicy and delicious;" the *cashew apple,* the "*mawa*;" the "*memgongo,*" a fruit much resembling a northern apple or pear; and the *mponko,* that resembles our melon.

Mangabas, goyabas, jambos, mammoons, cajas, cajus, cambocas, aracas, maracuja—these are some of the fruits of Brazil that to Northern ears convey no definite meaning, but which are rich, delicate and refreshing. Travelers in South America also speak of the *manirito,* the *madrona,* a delightful sub-acid fruit, and the *cacaita,* or monkey *cacao-bean.*

CLAY EATERS.

Not only organized but unorganized substances seem to be capable of sustaining life. Thus the Ottomac Indians of South America eat earth, which they swallow daily in large quantities, for months in succession. This earth is a fine unctuous clay, of a yellowish color, and when hardened has a little red color, due to the oxide of iron. During those seasons when fruits are scarce these Indians subsist almost entirely on this earth, which they make into balls and consume in enormous quantities, and take very little else besides.

Similarly the negroes on the coast of Guinea eat a yellow earth called *caauac.*

The inhabitants of Java eat cakes made of "reddish and somewhat ferruginous clay," but are injured by it and perceptibly lose their flesh. The Ottomac Indians seem to thrive on dirt-eating, for Humboldt describes them as "men of a very robust constitution," but they do not take much exercise.

The negroes of Gambia mingle clay with their rice.

The conclusions to which Humboldt arrives, are, that the habit of eating earth is confined to the tropics of both hemispheres; that the positive nutriment in these earths must be very small indeed, and that it acts rather to appease hunger, and stimulate the secretion of gastric juice ; that some tribes eat it from necessity, but others from choice.

It seems almost necessary to accept the view that these clays do contain positive nutriment which in some way becomes assimilated.

The General Law of Evolution is applied to Positive as well as to Negative Food.

In the general history I have given of the method by which men have arrived at their present knowledge and use of food, it follows that the law of evolution may be justly said to apply to food, and may explain and confirm its history. The history of food, like the history of stimulants and narcotics, like the history of all organic nature, is the history of a struggle for existence, of continuous differentiation, and multiplication, and adaptation to changes in soil, climate, and human progress.

In proportion as man has multiplied and advanced, and become subdivided into tribes and races, in that proportion has the food on which he has subsisted been multiplied in variety ; and in proportion as he has risen from savagery toward civilization, in that proportion has his food improved in quality as well as in quantity ; and side by side with modern refinement in the art of cookery there has been a great increase of nervousness and nervous diseases, that has made tempting diet a necessity.

All the vast institutions of mechanical skill, in the forms of trades and specialties, would have been hampered and retarded had not food correspondingly combined to meet the varying wants of all.

The early savages, we have seen, must have slowly or rapidly starved on barks, and leaves, and berries, with occasional feasts of meat, or fish, or fruit. The ancient civilizations rejoiced in tables that were then called abundant and luxurious, and at the banquets of the Roman nabobs all the known world was laid under explicit and extravagant contribution; but the Roman feasts were tame enough in comparison with the daily needs of the average European or American.

As man has migrated from land to land, and sea to sea, in that proportion he has found new, and different, and peculiar varieties of food adapted to his wants in the climates in which he has taken his abode.

Climates also have changed from age to age, and the materials of food which depend on climate have changed with it; and correspondingly, man's desires as well as his needs have also changed.

The evolution of food has kept place with the evolution not only of agriculture and commerce, but of social life, of religion, of mechanical skill, invention, and of morals. As man has cultivated more he has desired more; as his taste has become sensitive and capricious, the means to gratify that taste have arisen.

In the darkness of the Middle Ages, gluttons were abundant; but they indulged their passions with a limited and oftentimes untempting variety of food. For ages certain nations of Europe lived largely on pork, and their intellectual achievements well corresponded.

In America, the varieties of the raw material food are greater than in any other continent, and in addition, nearly all the desirable products of the world are brought to our shores.

THE LAW OF EVOLUTION APPLIES ALSO TO NEGATIVE FOOD.

Stimulants and narcotics have likewise multiplied and improved with the advance of the race, and are now used in their greatest variety in the most highly civilized nations.

To the *divine Saoma* has succeeded wine, and wine has differentiated into hundreds of varieties that the ancient world knew not of.

The opium and hemp of the ancients have been variously modified by modern cultivation, and are to a certain extent replaced by tobacco, coffee, tea, and chocolate, with their many imitations.*

In accordance with the general law of evolution, varieties of food that were formerly held in esteem, are now little used ; according as new and better varieties arise, more pleasing modifications are developed with the advance of the race. The gluttonous feasts of our forefathers would hardly be attractive to modern Londoners or Parisians.

Ages ago, sage was a common drink in Europe, but that, and numerous other herbs, have yielded to tea and coffee.

Already we have seen how rapid has been the evolution of cereals and fruits, and how many improved varieties are required, which in our fathers' day had not been born. The fruits and vegetables of our modern markets are as much richer and more varied, than those of the middle ages as the civilization of the 19th century

---

* For a more detailed description and history, see my work on *Stimulants and Narcotics*, of this series.

is more complex, and more refined, than the civilization of Mediæval Europe.

In the time of Henry VIII. there were no cabbages, carrots, nor any esculent roots in England, and about this time artichokes, apricots and plums first appeared.

Among the articles of food that were once popular, but now, for some reason, have lost their popularity, are the *lamprey*, once a great delicacy ; the *sturgeon*, of which the Greeks and Romans were so fond ; the *dolphin*, which is now regarded as insipid. The *porpoise*, once most popular, is now almost forgotten. Among birds the *bustard* and the *flamingo* were greatly lauded by the ancient Romans.

The majority of the inhabitants of the world are underfed.

Of the twelve hundred million inhabitants of this globe only a very small fraction systematically get enough to eat. The almost universal condition of existence is poverty, and in no other direction is poverty so widely felt as in want of food. Clothing of some kind, cheap or luxurious, and enough for health, is enjoyed by all races from the lower savages to the more enlightened. A suit when once provided lasts long, and when lost or worn out, can easily be replaced by some other, which is all that is needed for protection if not for luxury. There are very few, and those are usually only of the very poorest classes in civilized countries, who need to suffer from cold through lack of clothing.

Food, on the contrary, is usually more or less difficult to get, and civilization is always expensive. Even when food is abundant in quantity the quality may be so bad that those who subsist on it are poorly nourished. In this matter of diet extremes meet ; the savage goes hungry because he is too lazy to go after food, or, perhaps, because it cannot be found ; the wealthy and cultivated restrain their appetites because fashion—more tyranni- cal than hunger even—enjoins it ; or because they are so absorbed in intellectual pursuits that the cry of the passions is not heard.

The greatest gluttons may be chronically underfed, for we are told that the Siberians, the Kamtschatkans, and the African tribes and savages generally, who, at their feasts, appear to be enormous gluttons, go for days at a time without anything to eat ; and there are mil- lions and millions who live and die in this world with- out ever having known a really substantial and every way healthful meal.

Among enlightened nations only a small minority can afford to buy the best of food. The great mass must be content with what they can get, and cooked in a manner at once unscientific and unpalatable.

Carlyle said to Emerson that the best thing he ever heard of America was, that there every·one could have meat for his dinner—a remark which suggests very painfully the condition of the greater portion of the population of Europe—and yet of the 40,000,000 inhabi- tants of the United States—the most favored nation in the world in regard to food—there are probably not more than one million who can or do systematically live well.

In Great Britain—that has given the world the best civilization it has yet seen, and where among the higher classes the art of dining has been carried to its highest perfection—hundreds of thousands rise each morning knowing not where they shall get their breakfast, and there are very many millions who have never tasted and never will taste a generous and varied dinner.

In the United States the diet of the poorer classes in the great cities is so abominable that the physicians of the public institutions of charity find great difficulty in treating them—with satisfaction.   Our farming population, who raise their own food, live much worse than is supposed.   Pork, cabbage and potatoes are the great staples, and in thousands of families a piece of tender beef steak would be as much of a rarity as a *durion* of the Malay Archipelago.*

Thus we see that from the starving millions on the Plains of Persia, whom despotism and ignorance keep perpetually poor, and the bigoted castes of India dying by millions amid the cattle which their religion forbids them to eat ; and the lazy Siberian, too short-sighted to accumulate in seasons of plenty against seasons of scarcity ; and the wretched African tribes, decimated by disease brought on by life-long deprivation of the commonest necessities, to the gay lady of fashion, who goes hungry away from a bountiful table, lest she might become gross, or have blotches on her face, and the hard toiling brain-worker, who forgets whether he has dined in the excitement of his book or a problem, or whom crazy hygienists have taught that prolonged hunger is the true prescription for health and long life—through all grades of humanity inanition is the rule, repletion the exception.

Among the civilized portions of society of modern

* The statements on this topic, though necessarily given with considerable absoluteness, are not made at random.   They are based on careful study of those works that describe the manners and customs of different countries, and on the statistics of Europe and America.

times, the number of those who often overeat is very
limited. In the old civilizations of Persia and Rome,
and with our forefathers in Great Britain and Northern
Europe,* gluttony was a prominent if not universal vice
among those who could obtain the material on which to
be gluttonous. The vice is a relic of savagery, of a
social state when food was scarce, or at least uncertain,
and men were tempted to gorge themselves in order to
compensate for the famines of the past and anticipate
those that were to come. Then again, savages have so little intellectual culture and so few amusements, that the
gratification of appetite has limited restraints.

With the rise of modern culture, with the vast multiplication of books and newspapers, with the varied
opportunities for literary and social activity and re-
creation, in lectures, public worship, in political gather-
ings, in the multitudinous cares and details of science,
art, scholarship, education and trade, and with the varying and incessant demands of fashion, and, in connection with all these activities, with an improved moral
sense and consequent growing repugnance to coarse
vices, crimes of every kind are either disappearing, or,
retaining their confined essence, assume a more refined
character. Thus the love for bull fights and the gladi-
atorial arena, have given way to the pantomime, the
tragedy, and the ballet. Thus also in the best circles,
intemperate drinking, which only half a century ago
was almost universal in the leading classes of Great
Britain and the United States, and was held in esteem,
has now become among these classes an exceptional

---

* The word "gormand" is said to be derived from Gorman, a gluttonous
king of the Danes. The Normans were more temperate than the Britons, but
were won over to habits of gluttony by the people whom they conquered.

vice, and is held in great reproach.   Ardent spirits have,
to a certain extent, been replaced by the milder stimulants, tea, coffee, and chocolate.   Similarly, also, the disgusting habit of tobacco-chewing has slowly yielded to
tobacco-smoking, and carrying the refinement further
still, the filthy pipe has been cast aside in favor of the
beautiful meerschaum, the elegant cigar, or the light
cigarette.

All these habits, from the sturdy women of Queen
Bessie's reign gulping their beer by the quart, to the
fashionable lady of West End or Fifth Avenue, delicately sipping her weak chocolate or tea, have a common
basis, but the difference between the former and the latter represents the difference between the coarseness of
the sixteenth and the culture of the nineteenth centuries.

Civilization has vices enough, but they are different
from those of barbarism in their manifestation if not in
their purpose.   The corruption—which is the one great
vice of a high civilization—is simply a *refined* method
of robbery.   The man who in the 14th century would
have waylaid a stage coach and murdered the passen-
gers, now embezzles his employer's funds, steals from
the people by lying advertisements, and robs the treas-
uries of cities, nations, and railroads.

OVEREATING VERY RARELY A CAUSE OF CHRONIC DISEASES AMONG THE INTELLIGENT CLASSES.

Rarely, if ever, do I see a patient with chronic disease
that has been induced or aggravated by overeating.
Temporary disturbances in digestion are continually
caused by carelessness in the use of indigestible fruits
and other substances ; but *permanent* ill health, as a

result of excessive *eating of good food,* in cultivated, intelligent *adults,* is so rare that I do not remember ever to have seen a case.

The vast and increasing army of paralytics, dyspeptics, consumptives, nervously exhausted, and insane, is filled, not by those who overeat, but by those who *undereat.** The great difficulty that practitioners, of chronic diseases especially, encounter is in getting their patients to take sufficient nutriment. Very much of the benefit that follows traveling, change of residence to the seaside or the mountains, muscular exercise, electrical treatment, and various medicines, is due to, and is measured by, the increase of the appetite, and corresponding increase in digestive capacity.

Many of our most valued medicines—cod liver oil, cream, phosphates, and iron—act directly as food, and owe their value to their nutritive powers.

To restrain debilitated patients from eating in our days is as unscientific as it would be to bleed them a few ounces daily, as was formerly done in such cases, and would have a similar effect to increase their debility.

The most vigorous, independent, and energetic people in all parts of the world are flesh eaters, while those who live exclusively or mainly on fish and vegetables are inactive, phlegmatic, and stationary.

This conclusion is clear and inevitable from the facts of history, and well accords with our knowledge of comparative anatomy, and with common observation.

---

* I do not forget the well recognized disease, *boulimia,* which consists in abnormal and unappeasable appetite ; nor that epilepsy has for one of its symptoms or results a morbid desire for food that demands restraint.

Theoretically, it would seem that the flesh of lower animals, which so closely resembles that of man, would be of all food the easiest to assimilate, and the best form of nutrition. Experience demonstrates the correctness of this theory. The strength and working power of human beings is, to a certain extent, proportioned to the quantity of animal food that they can habitually digest.

The most powerful nations and the greatest and best men everywhere are flesh eaters. In all civilized countries the leading classes who control the civilization, have never been vegetarians, and so far as can be learned this statement is true for all the older nations—Assyria, Egypt, Greece and Rome; that it is true of all the great nations of modern times no one will question.*

Among the savage and semi-savage races we find that the energetic, the warlike, and the migratory, are flesh eaters, while those whose dependence is exclusively on fish, vegetables or fruit, are comparatively passive, timid, peaceable and stationary both in character and locality.† Contrast the wild and ugly tribes of Central Africa, who make murder a pastime, and keep their country half depopulated by cruel wars: the bold and adventurous herdsmen of the South American pampas; the fierce and wandering Arab of the desert; and the murderous aborigine of North America, and the well-to-do classes of civilized lands—all of whom are liberal or exclusive flesh eaters, with the

---

* It has been calculated that in France *one sixth* of a pound of meat is daily consumed ; the average for London and New York is about *half a pound.*

† I do not overlook although I have here no room to describe the elements of race, climate, government, religion and education, all of which more or less co-operate with diet in making nations what they are.

timid and oppressed Ryots of India, the peaceable Chinese and Japanese, stationary for centuries; the Laplanders and Icelanders, incredibly phlegmatic and indolent; the Siberians, the Kamtschatkans, the dwellers on the Pacific Islands, and the peasantry of civilization everywhere—all of whom depend wholly or largely on fish, vegetables and fruit, and we have facts sufficient from which to draw our generalization of the effect of flesh diet on the activity of the brain and muscle.

Pizarro found the Peruvians, who lived on vegetables, mild and peaceable, while Cortes found the same race in Mexico, living on flesh, fierce and savage.

THE LAW APPLIES TO THE LOWER ANIMALS.

The law is just as fixed in its application to the lower animals. It is stated the bears of India and America that live on acorns are mild and tamable, while those of polar climates, that feed on flesh, are fierce and untamable.

Even our domestic animals advertise their diet in their faces, and any one can tell a butcher's dog as far as he can see him.

Enough is known on this subject to enable us to know a priori, that the German tribes who overthrew the Roman Empire could not have been vegetarians or fish eaters, and that the great Germano-Scandinavians who colonized Great Britain and North America, and sent their commerce to every ocean, must have been a beef eating race.

CANNIBALS MORE COURAGEOUS AND ACTIVE THAN VEGETARIANS.

Human flesh seems to agree with those who eat it;

it is certainly far preferable to a purely vegetable diet. Du Chaillu says that the Fans—a cannibal tribe who eat even those who have died from sickness, and will steal bodies that have been long dead,—are "the finest, bravest-looking set" of negroes of the interior of Africa.

Tribes that devour human flesh are by no means, however, the most cruel, or the most degraded specimens of humanity. Cannibalism exists in different parts of the globe, and in very different races.

It is stated by Abd-allatif, physician of Bagdad, that in Egypt, in the 13th century, in a season of great scarcity, the habit of eating human flesh spread through all classes. At first it caused amazement and horror ; but the habit gradually extended until people of wealth and position regarded human flesh as a treat, and secreted it in stock against hard times. It was cooked in various ways. Of the poorer classes, large numbers were destroyed in this way.

Physicians were in especial demand. They were sent for on pretense of consultation and were seized and devoured.*

Reade, who has traveled among cannibals, expresses the same views. Says he, "A cannibal is not necessarily ferocious. He eats his fellow creatures not because he hates them, but because he likes them."

Even in the hottest climates meat seems to be required for those who toil hard.

GOUAMBA, OR MEAT-HUNGER.

Du Chaillu describes a disease very prevalent in

---

* This account I take from Humboldt's Travels. Bohn's ed., vol. ii. p. 416.

Equatorial Africa, called *gouamba.* It is an intense longing for meat, induced by long abstinence from it. It is very distressing ; one of Du Chaillu's companions, an African, suffered from it so much that he cried like a child. In that climate vegetables will not satisfy.

In this country those who, after living a long time on flesh, are suddenly and entirely deprived of it, are sometimes afflicted with this gouamba, at least in a mild form.

EFFECT OF FISH DIET ON THE INTELLECT.

The statement has recently gone forth on the authority of Prof. Agassiz and others, and has been very generally accepted, that fish diet, by virtue of the phosphorus which it contains, is pre-eminently adapted to nourish the brain, and that those who subsist on it largely are distinguished for their brightness and intellectuality. Now, while it is true that a small percentage of phosphorus enters into the composition of the healthy brain, and while it is also true that fish contains more or less of phosphorus, that may, and probably does, pass into the circulation, it is yet to be proved either by theory or by the experience of mankind, that a diet of fish is, *on the whole,* better adapted to supply the waste of the brain than a liberal variety of other alimentary substances, and especially of meats. Indeed, as we have seen, all the facts that in any way bear on the subject go to prove directly the contrary. It must be remembered that, besides phosphorus, the brain is composed of a number of other important substances, on a proper supply of which its activity and healthfulness materially depend, and there-that best supplies the waste of *all* those elements. Many of our common alimentary substances contain phosphorus, and the different varieties of meat contain it in about as large a percentage as fish. Pereira states that more phosphorus is afforded to the body than it requires when flesh, bread, fruit, and husks of grain are used for food, and the excess is eliminated from the system.

Furthermore, the different varieties of meat, besides containing nearly as much phosphorus as fish, have a greater percentage of nitrogeneous substances, and a much less quantity of water; and must therefore, necessarily, be more nutritive for the brain as well as for the system generally. It is found by universal experience that fish is not only less satisfying to the appetite than meat, but that it is also less sustaining to the mental and physical strength. Herodotus tells us that certain classes among the Babylonians and ancient Egyptians subsisted mainly on a diet of fish, and we are led to believe that they belonged to the inferior orders of society. The largest and most exclusive consumers of fish in the world are the Chinese, the Hindoos, the Japanese, the savages of New Zealand, the Siberians, the Icelanders, and Laplanders, the Indians of the Amazon, the Greenlanders and Esquimaux, and nearly all the dwellers on the northern coasts of Europe and America. The strongest admirers of the Japanese will surely not presume to compare them with the beef-eating Anglo-Saxons. The intellectual inferiority of the Greenlanders, the Esquimaux, and the natives of New Zealand, will be conceded without argument.

If maritime nations have exhibited greater vigor and ability than the inhabitants of the interior, as suggested

Prof. Agassiz, the cause is to be found not so much any peculiarity of their diet as in the recognized fact history, that the superior races have intuitively ished their way to the sea-coast on account of the portunities there afforded for commerce and colonition.

To all this it may be added that the greatest fish ters are not always those who live near the sea. ibes who live on rivers subsist on fish, even in the terior of Africa and America.

The Icelanders are notoriously indolent and phlegatic, and stock-fish is their daily food. Travelers all ree in these two statements, that these people have h on their tables nearly every day, if not every meal; d that all the inhabitants exhibit a lethargy, a cold- ss, and a slowness of movement that to the beef- ting English is astounding. Making allowance for e climate in which they live, there is much in their aracter, as well as in that of the Esquimaux, that can- t well be accounted for by any other theory than that h must affect them unfavorably.

The character of the Chinese and the Hindoos for ssivity and for quietness is sufficiently known, and of e Siberian fish eaters it is enough to say that they liberately prefer starvation to work. Terrible famines cur among them because in seasons of abundance they nnot muster the energy to store up the salted fish, iich is their chief subsistence, against seasons of arcity. The aborigines of the Amazon—whose great pport is fish—are described by Wallace as not only aceable but bashful. "They scarcely ever quarrel iong themselves, work hard, and submit willingly to thority." This description contrasts powerfully with what African travelers, as Walker, Livingstone, Speke and Du Chaillu give us of carnivorous tribes of that equatorial continent.

Civilization is indeed but little, if any, more indebted to fish than to vegetables and fruits.

FISH EATING COMMUNITIES INTELLECTUALLY INFERIOR.

The operation of this law is just as clearly seen when it is studied in its application to isolated communities.

The stupidity and indolence of fishermen and fishing districts is proverbial. It is necessary to go no more than an hour's distance from New York—the most active city on this continent—along the coasts of Long Island and New Jersey, to find arguments in favor of the stupefying effect of a fish diet.

The inhabitants of Cape Cod and the adjacent islands, Nantucket and Martha's Vineyard, are probably the most exclusive fish eaters on this continent; besides all the abundant varieties of salt water fish along the coast and the bays, and the finest of trout in the brooks, herrings, rightly termed "Cape Cod turkey," are caught by thousands, and are dried and smoked, and on many tables are never absent during the three hundred and sixty-five days of the year. Fresh beef is scarce; farming is poor enough; poultry is not over-abundant, and on their sandy soil no more than ordinary vegetables or fruits can be produced. Those who have the sunniest memories of the unsurpassed hospitality of the people of the Cape will surely not claim for them any more than average intellectuality, but must admit that in this respect they compare unfavorably with the bold and energetic Yankee stock of which they are a part.

Probably no class in American society have given so few great men to the country as fishermen.

FISH DIET A SOPORIFIC.

All the evidence of experience goes to show that fish not only does not increase the activity of the intellect, but that its effects, both temporary and permanent, *are distinctly the reverse.* It acts like a *soporific*, or *sedative.*

Those who suddenly change from a generous mixed diet to one where fish takes the place of meat, oftentimes observe that they become sleepy and stupid, during the day ; a sort of loginess comes over them very much like that which those experience who indulge freely in milk or beer. This effect is by no means constant ; but it is not unfrequent. I have frequently observed it in my own personal experience, and have seen many who have made the same observation on themselves.

An intelligent patient of mine, an athlete, who has given considerable attention to the diet of gymnasts and muscle men, says that he once subsisted mainly on fish for ten weeks, living in all other respects as usual. He was troubled with drowsiness and stupor that made it hard for him to go through his daily duties. When fish diet is used pretty exclusively, these effects become a fixed condition ; and therefore we find that fishermen everywhere, and nations who, like the Icelanders, subsist on fish, are sleepy, indolent and phlegmatic.

What chemical principle in fish it is that produces this calming effect I cannot say. Possibly it may be the phosphorus, for the phosphates when given as medicine do sometimes aid the sleep, and are therefore prescribed for insomnia.

To a far less degree fish acts like bromide of potassium—a remedy which, when given in large doses for months and years, is, by some physicians, believed to weaken the intellectual force.

Like bromide of potassium, fish may be of service to a tired brain, not by making it work more, but by making it work less, by compelling it to rest, and therefore it may be well to spend a part of our vacations where the wheels of thought should stop, or at least go very slowly, where we can diminish the life current by an impoverishing and calming diet of fish caught by our own hand.

On the same principle it may be well for this nervous, brain-working generation, to learn lessons of hygienic wisdom from Iceland and Norway, from China and Hindostan, and give the beef of the waters a more prominent position on our tables, not in order that we may do more, but that we may do less.

The external attractiveness or repulsiveness of the different forms of animated nature that are used as food, stand in no relation to their palatability and value as nutriment.

Sight and smell, and touch, and hearing, are all deceptive guides in the selection of food.

The durion is one of the richest and most luscious fruits of the tropics, but its odor is so disgusting that it drives those unaccustomed to it from the room, or even from the neighborhood in which it lies.

The opinion that animals which feed on repulsive diet are therefore unfit for human food is baseless. Life s a wonderful chemist, and in its hidden laboratories ransforms the most nauseous and unsightly food into weet and healthful flesh.

Carnivorous birds, vultures, etc., are not esteemed as ood, but other animals, as pigs and domestic fowls, vhose flesh is so valued, are as filthy in their diet as can e imagined. Wallace, who has written the best works f travel since Humboldt's day, says that " carnivorous sh aré not less delicate eating than herbivorous nes."

In Guatemala there is a belief that lizards eaten alive ure the cancer.

That frogs are a luxury we all know, but it is said hat toads are often taken to market as substitutes, and ive entire satisfaction. The natives of Surinam and)ominica certainly eat the toad and make soup of it for he sick. The iguana is hideous enough, but the na- .ve of Jamaica would feel much wronged if he were eprived of it.

Beetles were eaten by the Roman epicures in order to itten themselves, and the Turkish women, it is stated,)r the same purpose eat beetles in butter.

In Sweden ants are distilled with rye for brandy to ive it a flavor.

In Ceylon bees are eaten ; in Australia the moth a great delicacy ; among the Hottentots cater- llars are as much of a luxury as sugar plums are ith us.

Spiders have been craved by some intelligent and ltivated people, as the stories go, and the Bushman gards them as a dainty. Centipedes also, Humboldt clares, are swallowed with avidity by the children of)uth America.

In 1542 a cloud of locusts invaded Germany ; they were broiled on a gridiron and found to be wholesome ; thus they were the means of saving the lives that they intended to destroy.

The armadillo and the porcupine are eaten in South America, and said to be delicious even to Europeans.

The alligator is far from being a favorite when alive, but when well broiled its flavor is not unlike that of the lobster, and makes excellent food.

The electric eel, though dreaded by man, in the region where it abounds is, by gourmands, highly rated.

The Llaneros of South America prefer to eat turtles in darkness, or the reason that the choicest bits would not be eaten if seen.

The smell of the carcass of the hyena, is so abominable that even the dogs turn away from it, but the Arabs manage to eat it with voracity and pleasure.

Some of our most highly rated articles of food must have long gone untouched on account of their repulsiveness.

Brave must have been the man who ate the first shrimp, crab, frog, or oyster.

Says the poet Gay—

> "The man had sure a palate covered o'er
> With steel or brass, that on the rocky shore
> First ope'd the oozy oyster's pearly throat
> And risked the living morsel down his throat."

Skunk is called better than pork, even in the United States, by those who can get near enough to him to catch him.

[uc, describing a Tartar banquet, says that the intes-
es, liver, and heart of a sheep were boiled in a pot
a some condiment; "the intestines being preserved
ire, and arranged as they were in the stomach of the
nal."

Vhen we dress an animal, we take out the entrails
. eat the body; but there are millions who regard
entrails as the greatest luxury, and if food be abun-
t, throw away the rest of the animal.

*'ood which is at first repulsive and disgusting to the ap-
te, may, by frequent use, very readily become agreeable
: useful, and this acquired taste may be transmitted.*

Jot only the senses of sight, smell, and hearing, and
ch, but even the appetite itself—which is the best
de we have in the selection of our food—is by no
ans infallible. The system may be nourished and
l nourished by substances that shock the palate and
iseate the stomach.

Travelers agree that many of the hideous articles of
d in use among the savages, and the very thought of
ich causes fainting and sickness, may in a short time
ome most delightful and satisfying.

t is not necessary to travel to demonstrate this pro-
ition; for do we not see tomatoes on all our tables,
l only a minority like them at first? To almost every
they are disgusting, but after long practice has
rcome this disgust, they become a most agreeable
ary.

APPETITE HEREDITARY.

!verything in the human organization, good and bad,
iibject to the great law of inheritance, and with the

reversion and other modifications which are a part of the general law, are liable to be transmitted from parents to children.

By this law we may explain the facts already presented, of whole nations who eat with relish varieties of food that to other nations are so repulsive; that decomposed fish and flesh, sour milk, are by so many preferred to these articles in a fresh and pure condition.

It is more than probable that twenty five years hence children will love tomatoes as naturally as they now love peaches or potatoes. It is hardly probable that man, even in his lowest state, originally preferred putrid and decomposed to fresh nutriment. The theory is more plausible that on account of scarcity or negligence, they were often obliged to eat food that had passed into decomposition, and that by long use they learned to prefer that flavor which at first was disagreeable, and that this taste was transmitted to posterity.

This theory is re-enforced by the fact that it is mostly among the savage tribes, or the wretchedly poor, that these strange tastes are found, and that under increased comforts they have mostly disappeared. In the same way I account for the preference which the Icelanders and the Irish peasantry give for sour milk; for such preference, like the preference for high game, is but exceptionally seen among the intelligent and well-favored classes of any country.

Not only taste but probably also distaste is hereditary. With enlarged resources and refined cookery, enlightened man has gradually abandoned articles of food that were once his favorites, and has lost his love for them, and it is surely not irrational to infer that this dislike has been inherited, so that we now almost tremble at the

ght of sitting down at the tables of our ances-
*

*arly all people reject, even under the pressure of
me necessity, some one or many articles of food at
command, and the causes for such rejection are often-
strange and unaccountable.*

perstition and religion have much to do with the
tion of many articles of food, especially among cer-
savage nations.

ie Kaffir, like the Jew, will never eat pork, and he
tins also from fish ; and yet he seeks the raw flesh
ie ox, even the most disgusting portions.

ie Pacific Islanders, like many other savages, are
ented by their religion from eating many valuable
les of food.   One worships his god in the shark ;
her in the eel ; another in the owl ; and so on
igh all animate nature.

ie East Africans, though very fond of animal food,
ie eggs, and consider cheese a mineral, and there-
denounce it.

daism prohibited pork, as all know, and the religions
idia compel vast millions to practice absolute or
oximate vegetarianism.

en as late as the middle of the nineteenth century

---

ose who desire to learn by experience how easy it is, in a comparatively
ime, to acquire a taste for eating rank food, and a corresponding distaste
t which is pure, may have a good opportunity on long voyages.

s my misfortune, during the late war, to spend fifteen months on board
of the steamers engaged in blockading the ports of the Western Gulf

For nine months we lay anchored without once going ashore.   Though
ell supplied with canned and salted provisions, the butter was almost al-
ncid and hideous to the taste, and the odor was so powerful that, when
proposed to slush the masts with it the paymaster protested, declaring
it were done " the blockade runners could smell us three miles off."
urning North the sweet, fresh butter, was positively insipid.   The ex-

we have seen a not inconsiderable number of Protestant Christians abstaining from wine for conscience sake. On the festal days of the Roman, and especially of the Greek Church, flesh of all kinds is prohibited.

With the general advance of the race, and the evolution of freedom in thought and action, the interference of religion with diet has been gradually reduced to a minimum.

The Llaneros, a pastoral people of South America, despise milk and butter, regarding them as fit only for children.

In the United States we call rabbit a delicacy, but a negro of the West Indies will go very long hungry before he will touch it.

As a general law all varieties of food are best adapted for the climate in which they are produced ; but this law is susceptible of very wide modifications.

That the train oil and blubber of Greenland are better adapted for those regions than for the temperate zone ; that there is a certain correspondence between the fruits of the tropics and the demands of the system in those regions ; that those travelers best succeed in preserving health and strength who measurably conform to the eating and drinking customs of the countries they visit—all these statements are but truisms, and need no detailed exhibition of authorities to make them acceptable.

And yet the history of the world, and especially since the recent and rapid development of modern commerce, shows that civilized man can use with profit and pleasure, food gathered from all the ends of the earth. It appears not to be well to live entirely or mainly on im-

1 food ; and none long try the experiment. The
itants of the temperate zones are the most for-
e in this respect, since they can draw from the
c zone on the one hand, and tropical on the other;
probably without necessary detriment to health.
ere is no satisfactory evidence that the fruits and
ments brought to us from the tropics have any
than a pleasurable and beneficial effect, when
with reasonable caution and with a due regard to
idual idiosyncrasies ; but the desire for these sub-
es is much less active in the temperate zone than
e countries where they are produced.
n seems to demand and to bear wider extremes in
od than the lower animals ; for we all know that
tropical birds and polar animals die readily if
pains be not taken with their nutrition.
follows from this that the products of the tempe-
zones should have a wider applicability than the
icts which can be raised only in the tropical or
regions.
e inference is confirmed by experience.
tatoes, rice and the cereals, are valuable food in
climate to which they may be carried.
nilarly with negative food. The most universal
ilants are tea and tobacco, and both are the pro-
of temperate climates.*

*tremely hot and extremely cold climates demand oily
atty food, or other carbo-hydrates, though perhaps not
ial quantity.*
big, the father of physiological chemistry, has made
ision of food that is as familiar as it is unsound.

According to his classification the heat producers will
least of all be needed in hot climates. Experience shows
that they supply a want in the deserts of Africa as much
as amid the snows of Siberia.

Raw walrus beef, Dr. Kane declares, is the best fuel
a man can swallow; but what shall we say of the travelers in the deserts of Persia and Arabia; of the herdsmen of the Pampas of South America, who, under a
burning sun, feel similar if not equal need of oil and
fat?

No other facts than these are needed to show the
difficulty and present impossibility of solving the complex problem of diet by physiological chemistry.

Baker tells us of an Arab sheik, who, though
"upwards of eighty, as upright as a dart, a perfect
Hercules," had daily consumed through his life two
pounds of melted butter.

Baker further states that "fat is the great desideratum
of the Arab." Not only does he smear his beard, his
clothes and his body with it, but he eats all he can get.

Du Chaillu states that the Bakalai tribe of Africa
have more disease than the Apingi, and he attributes
this difference to the fact that the latter tribe consume
more palm oil, which, in the absence of game, constitutes
a considerable portion of their food. He further states
that this tribe are exceedingly fertile as compared with
other African tribes, and raises the query whether the
immense amount of palm oil they eat may not have
something to do with it.

Dr. Livingstone, speaking of the borders of the Kala-
hari desert, says:

"A considerable proportion of animal diet seems requisite here. Independent of the want of salt, we

ire meat in as large quantities, daily, as we do in
land, and no bad effects, in the way of biliousness,
w the free use of flesh, as in other hot climates. A
table diet causes acidity and heartburn.

irton says that in East Africa meat is the diet most
d, and that those who can afford it live on flesh
st entirely, considering "fat the essential element
od living."

e states, furthermore, that although fish are abun-
in the rivers, they are despised by those who can
lesh.

ie Persians live on rice and oil.

Central and Eastern Arabia the date is the food of
Arab, the staff of his life, and chief commercial
uct. Mahomet knew its value when he com-
led: "Honor the date tree, for she is your
ier."

ie Arab of the Desert of Sahara drinks oil as we
x coffee. When it is in abundance he can sustain
; fatigue. Says Richardson: "An Arab will live
months on barley meal paste dipped in olive

would seem that the different varieties of the carbo-
ates are, to a certain degree, substitutes for each
.. In India and China the leading carbo-hydrate is
in Central and Eastern Arabia, the date; in the De-
if Sahara, and in Eastern and Western Africa, Syria
Palestine, palm and olive oil, and in nearly all those
ries milk and honey, and liquid butter, cocoanut
ground-nut oil, and fat and lean meat of various
are used.

East Africa, where the cucumber grows wild, an

the methods by which the inhabitants of extremely hot regions satisfy the demand for carbo-hydrates.

In East Africa the people chew the sugar cane, and "sugar attracts them like flies; they clap their hands with delight at the taste; they buy it for its weight in ivory; and if a thimbleful of the powder happen to fall upon the ground they will eat an ounce of earth rather than lose a grain of it."

The Mussulmans of the desert cook their rice in fat.

In Chili balls of grease are sold in the markets, and in Egypt nearly everything is cooked swimming in fat, and in Madagascar traveling companies say that grease is mixed with all their food.

In Chili also sugar is much eaten.

Hot oil as well as hot gravy is taken by the Arabians of the desert, and is given as a medicine in certain diseases. Thomson says that the principal food of the inhabitants of Syria and Palestine is cheese, figs, olives and sour milk, and that almost every dish is cooked in oil.

In hot countries, where meat is scarce, or where it is prohibited by religion, other substances, classed by Liebig as heating foods, are used. The nutritive power of oils and fats is several times greater than that of rice, and therefore, when rice is used without oil or sweets, large quantities are required. The Hindoo eats enormous quantities of rice and butter, eked out sometimes by a little flesh; but at best this food is an imperfect substitute for meat or oil, for we are told that the quantity required for those who live mainly on rice is enormous, and that the poor Ryots, in consequence, become "pot

bellied," and present an appearance of chronic starvation.

On the contrary, the Arabs, who live on abundance of
flesh of various kinds, and oil and butter, and some of
the tribes of Central Africa, whose diet is similar, though
varied by human flesh, as well as the herdsmen of South
America, who rarely need to suffer for want of meat or
oil, are, for savages, vigorous, active, energetic and cour-
ageous.

Extremely hot regions not only require meat and oily
food, but enormous quantities of it ; probably not much
less in some sections than is required in regions of extreme cold.

Of the gastronomic performances of the Hottentots
every one has heard.

Barrow states that they are prodigious gluttons. An
ox of medium size was disposed of by ten Hottentots in
three days.

Probably more food is, on the average, eaten in extremely cold than in extremely hot climates, but there is
reason for believing that the difference is less than is
generally supposed. Like the Hottentots, the African
cannibals, and the Arabs, the Siberians and Greenlanders appear to be greater gluttons than they really are,
because they are compelled often to go hungry for many
days at a time. It is certain that the Laplanders,
the Norwegians, and Icelanders, are not remarkable
eaters.

The traveler, Vambéry, says that the Persian pil-
grims whom he met on his travels, were gigantic
eaters.

The effects of extreme heat and cold are analogous
in other respects.

"The drought and heat of the Llanos act like cold upon animals and plants," says Humboldt. In the season of extreme heat, boa constrictors and crocodiles remain torpid for weeks in the mud. Travelers in Africa confirm this observation.

Certain diseases, as leprosy and elephantiasis, seem to be common to tropical and polar regions. They are found in Iceland, and in Africa and Arabia. Ice and hot water often have a similarly curative influence when locally applied for the treatment of disease.

There are large numbers of people who prefer their flesh and fish and other food, when it has passed into a state of decomposition or fermentation.

In Burmah, the *gneper*, a huge fish, is eaten, sometimes in a semi-putrid or semi-pickled state.

The herdsmen of South America seem to prefer decomposed meat, even when the fresh is abundant on every hand.

The inhabitants along the Senegal and Orange rivers, prefer decomposed fish to those just taken out of the water.

Even in Sweden stock fish steeped in a solution of potash until decomposition takes place, is for the Christian's dinner what roast beef is to the Englishman.

The herdsmen of the Pampas of South America eat decomposed oxen with great relish.

There are some few among us who prefer old and rancid cheese to that which is fresh and newly made.

Even in civilization the old taste lingers; traces of barbarism are not yet wholly worn out; for even now there are those who prefer their game high.

Fermented liquors are enjoyed by the most enlightened equally with the most degraded ; but fermentation
is a species of decomposition.

Putrefaction and fermentation are seen to make some
alimentary substances more healthful, and to remove
injurious qualities.

The juice of the mandioca is at first poisonous, until
fermentation has commenced. Sour milk seems to be
less likely to do harm than sweet milk ; at least it is so
believed among certain tribes of Africa ; and the
world over, there are more who drink sour milk
than who take it fresh from the cow.

*Much of the disease of the world is caused by the
food.*

Between savagery and civilization, through all the
grades, this law applies ; and that the scurvy and
horrid diseases of the skin, and general low condition of many of the savage tribes are caused by
their abominable food, travelers well agree. The
delusion that all the diseases of humanity are confined to civilization is quickly dispelled by a careful
reading of works of travel.

Scurvy is caused by want of fresh vegetables and
fruits, and is removed by abundance of these articles.

The raw meat of the Abyssinians, when very freely
eaten, causes, it is stated, a kind of insensibility, not
unlike intoxication.

Almost the whole Abyssinian people are afflicted with
tape worms ; it is their most prevalent disease; and
is without question due to the *raw* meat on which they
subsist.

The natives of Polynesia sometimes eat sharks, and

gorge themselves to such an extent that they vomit.
Their flesh is dry and acrid.

It is thought that the excessive and disproportionate
use of the date among the Arabs is a cause of indigestion.   An exclusive fish diet seems to cause various
disorders of the skin.

*The experience of mankind in the selection and com-
bination of food, is, to a certain extent, explained and confirmed by the sciences of organic and physiological chemistry.*

If man were forced to depend on his knowledge of
the relation of minute chemistry of food to the human
body, he would starve before he could prepare a single
meal; and after the utmost care and skill, in which the
most advanced science should be brought into requisi-
tion, he could not tell whether the first mouthful might
not instantly throw him into fatal convulsions.

But after experience has indicated to us the food
that we need, chemistry and physiology come in very
appropriately, to explain, in a most interesting manner,
the laws and principles thus ascertained, and to guide
in their application.

We cook some varieties of fish in oil; others, which
contain sufficient fat, are eaten alone.

In our puddings, eggs, milk, suet, and butter are
mingled with rice, and crackers, and bread, and tapioca.

In our salads are mingled oil and eggs, with lettuce
and chicken.   Rice is boiled with milk, and cheese is
eaten with macaroni, and green corn needs the addition
of butter and salt.

Buckwheat cakes are eaten with butter, milk, sugar,
molasses, honey, gravy and meats.   Everywhere vegeta-

bles are eaten with meats and butter with bread, and crackers with cheese. Pork and beans is a union of opposites as scientific as it is popular.

A modern dinner, beginning with soup, and ending after various courses with fruits, is, in the main, consistent with what little is known of physiological chemistry. Soup is an excellent preparation for more solid food, and raw oysters taken early in the meal serve to stimulate rather than deaden the appetite.

CHAPTER V.

DIETETICS is at once a science and an art.

It is a science, since it embraces a knowledge of the nature and the effects of food, as variously modified by soil, climate, race, season, social condition, and by methods of perfection and combination. It is an art, since it embraces the practical application of what is known of food and its uses.

Like most other sciences and arts, it is yet incomplete and is progressing.

Although it is known but in part, yet it is known much better than it was centuries, or even one century ago ; and we have reason for the conviction that it will be still better known in the ages to come.

Dietetics, like theology and like medicine, has advanced in waves rather than in a steady tide ; between the periods of advance there have been periods of depression.

Like many other sciences, also, dietetics has experienced violent reactions and revivals. Periods of enervating excess and luxury have been followed by periods of still more enervating asceticism.

Mr. Froude has shown that Calvinism was a protest against an age of self-indulgence and vice, a reaction that was necessary to recall man to his duty. Similarly, Vegetarianism and Grahamism, that were at one time so notorious in England and America, were reactions from the excessive indulgence, and the sensuality to which the Anglo-Saxons had for centuries been addicted. England and America were peopled by a race possessed of a strong passion for food and drink such as perhaps no other race of the world has ever exhibited. So far back as the records carry us they were pre-eminently a food-loving and alcohol-loving race.* No other people—not even excepting the wealthy Romans—took such delight in their feasts, or mingled their eating with such boisterousness and vulgarity. A race whose gods spent their time in rioting and feasting, and in the exercise of carnal pleasures, would not unnaturally regard a capacious stomach with more reverence than a capacious brain, and would estimate their friends by the amount of·liquor they could take at a time.

From all this sensuality came, both in England and in New England, a powerful reaction which, even now, has not found its level. What Puritanism was to religion, what the bleeding and calomel treatment was to medicine, what the rigid and oftentimes cruel common school system was to education, such was vegetarianism to dietetics.

Among the principal laws of diet that have been established by experience, and more or less explained and confirmed by chemical and physiological science, are these :

---

* See my work on " *Stimulants and Narcotics* " of this series

1. *The appetite, with all its liabilities to error, is unde*
the guidance of individual experience, the best guide u
have in determining the quality and the quantity of ou
food.

2. *The food of all classes should be generous, both i*
quantity and quality; and there should be an agreeabi
variety, not only from week to week and day to day, but (
each principal meal.

3. *The quantity, the quality, and the variety of foo*
must be modified by race, climate, season, age, sex, occu
pation, habit, and the progress of civilization.

4. *The principal meals should be taken with tolerabi*
regularity, and in calmness and leisure, and amid agree
able social surroundings.

The *first* law is so natural that we may well wonde
that it has taken us so long to find it out. If the objec
of food be to supply the waste of the system ; if withou
this supply the body grows thin and weak, and in a fer
days must die ; if the amount of work that the bod
can perform is proportioned to the food that is digested
and if the object of appetite is to inform us that th
system needs food; then surely it must be, on the whole
with all its liabilities to disease and error, the bes
guide we have in determining what, and how much w
should eat.

No one should long go hungry. We should rise fror
the principal meals with a feeling, not of hunger or de
sire for more, or of satiety, but of satisfaction.

Hunger is the cry of the body for food ; and it shoul
be answered with promptness and satisfaction.

DIFFICULTIES OF PHYSIOLOGICAL CHEMISTRY.

Whoever professes to apportion our diet by the scales, must first tell us just what and how many changes of tissue take place in the brain in every thought evolved, just what each idea costs us in elements of phosphorus, fat, and salts. He must tell us just how much fuel is consumed for every process of reasoning and every heat of passion. He must tell us how much nervous power is expended when the will commands a muscle to move, and how much protoplasm is transformed when the muscle obeys. He must keep an accurate record of every muscular contraction, and understand what number take place in the four hundred muscles of the body. He must rigidly count every beating of the heart, and every breath we draw. He must estimate and pursue every glance of the eye, and every change of the countenance. He must photograph each image that strikes the retina, and record every sound that falls upon the ear. He must weigh the imagination in scales, and the emotions in a balance. He must allow for every flight of fancy, for every glance of joy, for every weight of sorrow. He must gauge our very reveries, and as dreams are attended with changes of tissue, he must enter into even our night-watches, and catching our visions as they fly, must tell us what they cost the brain by the table of avoirdupois ; but when he has done all this his duties have just begun, for except he know the secretions and excretions, the other gigantic labors would be valueless. He must measure the blood as it pours through the arteries, and count the globules that are borne in its tide. He should know the product of every gland and the peculiar secretion of every organ. He must be able to state precisely the quantity of gastric juice in the stomach, and of bile secreted by the liver. He must collect and measure the watery vapor excreted by the lungs and skin, and all the waste products of the body. He must intimately know the actual condition of every molecule of the system, whether it is in a state of health or disease. If the body or any portion of it is in any morbid phase, he must accurately know the seat of the disease and all the modifications that it produces on the system; and when he has arranged this one side of the equation, there yet remains the difficult task of completing it by reducing to their last analysis the elements of his food.

If any man were endowed with divine power to know all these complex factors, it is manifest that unless a great miracle were wrought to enable him to make his calculations, he must starve before he could prepare a single meal.

If the time ever arrives when physiological chemistry is an exact science, so that by a mathematical computation of the changes of tissue that take place in the body, we can determine precisely the amount of nutriment that the system demands at each meal, it may be possible to apportion our meals by the scales, as is done with prisoners and paupers. Even though it were easy to regulate our diet by the rule of three, yet regard for comfort, and enjoyment, and economy of time, would require us to fall back on the natural appetite as the last and best appeal.

Appetite may become perverted and diseased so that it imperfectly represents the needs of the system, and the powers of digestion; and, as I have shown, it may be educated to enjoy what is at first repulsive. But with all its liability to err, it is, on the whole, a safer guide than any system of weights and measures, or any arbitrary laws. Those who always consult what they believe to be their rational appetite, will sometimes, perhaps frequently, go wrong, judging by standard of absolute correctness, and at times, may commit very grave errors; but in the end, and on the average, they will make fewer mistakes, and achieve higher success and usefulness than those who attempt to regulate the quantity and quality of their nutriment by chemistry, or mathematics, or by the authority of eminent names.

We are told to set aside, before a meal, just what we determine to eat, and confine ourselves strictly to that and no more. Such a habit will assist one to develop self-denial and force of will—the divine art of saying *no*—but will be likely to injure the health. We cannot tell beforehand just what we need for a meal. Appetite often grows as we eat, and it is proper that it should. The amount that we eat should to a certain extent be regulated by its palatability, but we cannot tell how our food will taste until we have tried it. Then again we often come to our meals so wearied and worn, by mental or other labor, that we are not in a condition to judge of the amount we need, until the meal is half over.

We are told to rise from the table as hungry as when we sat down. This direction, which has been made from time to time, by popular charlatans in hygiene, derives its plausibility from the fact that we are better able to study after a light than after a heavy meal. But those who proclaim this theory, forget that man was not made alone to study. He was made to eat, to drink and to

digest, and to enjoy his eating, drinking and digesting
in order that the power expended by activity may b
properly supplied.   The process of digestion costs some
thing.   It is itself a draft on the strength.   It increase
the very loss of tissue that it is designed to restore
When the digestive organs are working under full pres
sure, there is less strength and less disposition for men
tal labor than when the stomach is partly or wholly
empty.   But this is no reason why we should go hungry
An engineer who should keep his engine but half sup
plied with fuel for fear that it might wear out too fast
would be regarded as a lunatic.   And yet those who
systematically starve themselves, in order to rest the
organs of digestion, are even more unwise than the
engineer, because the stomach, and indeed the whole
system, is injured and weakened by underfeeding.   The
body can go longer without a supply of food than the
engine ; but with both it is a question of time merely,
and sooner or later, if the water is not supplied, the
wheels must stop.

We are told to avoid fine bread, to confine ourselves
to that which is unbolted and coarse, and that in Grahamism lies the solution of the great hygienic problem
of civilization.   Now while it is indubitably true that
unbolted flour is in some respects superior to that which
is bolted, yet that superiority is no reason why those
should use it who do not like it, much less depend on it
exclusively.   In all these matters nature is wiser than
the scientists.

It is a sad and suggestive fact that those who were
most led astray by these pernicious heresies were men
of the purest ambition and ripest culture—men who in
other departments of thought were sensible and well-

informed.   Coarse, unthinking natures, with stomachs of iron and brains of wood, usually care little for hygiene ; on the other hand the finely organized, impressible temperaments, who are rendered vigilant by capricious digestion and are continually susceptible to nervous derangement, whose delicate machinery only runs well when in perfect order, and is perpetually liable to become disturbed, are compelled by the very necessities of their being, to study the laws of health.   In default of proper instruction and guidance they are easily led astray.   Their love of inquiry becomes to them a source of error.   If they were less able they would make fewer mistakes.   All this comes from the ignorance of the principles of hygienic science, and from the lack of scientific method of teaching in our schools and colleges. If the very excellent men who so zealously espoused the cause of vegetarianism in the days of its rising popularity had been even moderately well taught in the true philosophy of diet, if they had ever been trained to scientific modes of thought—to the careful balancing of opposing facts, and to dispassioned, accurate observation—they would readily have seen the theoretical folly and practical absurdity of the views for which they so stoutly contended, and in defense of which some of them laid down their lives.

The history of the rise and fall of vegetarianism in America is exceedingly suggestive.   At one time it was espoused by some of the ablest minds in our colleges. It became a mania.   It was believed that indulgence in meat was the prime cause of the ill-health of students, for which the only antidote was vegetarianism. It was claimed that those who dined on potatoes and unleavened bread, were repaid for their self-denial by

greater keenness of intellect and stronger powers of application ; that they could go from the table to their
books with an ease and elasticity to which eaters of flesh
were utter strangers.   But abstinence from meat was
not by all regarded as self-denial.   In the excess of
their enthusiasm the disciples of vegetarianism declared
that simple fruits and bread were as much more palatable than unclean meats, as they were less injurious to
the health and morals ; that they partook of their hum-
ble meals amid joys which the luxurious flesh-eaters
knew not of.   I am not here describing a prehistoric
age, nor an ancient civilization, nor the days of our forefathers.   I write in memory of the time when the teach-
ings of the late Dr. Alcott, one of the most deluded and de-
luding writers that ever enlisted in the service of Hygiene
—whose amazing absurdities were only matched by the
ignorant sincerity and misguided industry with which
they were advocated—exerted a strong and practical
influence in many of our seats of learning.   I write in
memory of the time when the celebrated work of President Hitchcock—"*Dyspepsia Forestalled and Resisted*,"
one of the most unfortunate books on Hygiene that ever
proceeded from the pen of man,* was the law and the
gospel on the subject of diet, to a large number of in-
quiring minds in our academies, colleges and theologi-
cal seminaries—when the influence of the erratic Graham
was perceptibly and painfully felt wherever there were
in our land those who gave time and thought to the
subject of health.

---

* It is but justice to the memory of a good man to say that at the time when
President Hitchcock's work first appeared, some of the views which it contained were received by men in the profession.

CAUSE OF THE VEGETARIAN DELUSION.

And just here may I ask, what is the origin of the wide-spread delusion that meat is more difficult of digestion than vegetables? It is probably due in part to the influence of the vegetarian reformers, and partly also to the fact of every day observation, that salted meats—such as rare bacon and fried pork, and tough corned beef—are very trying to a weak stomach, and bring on many evil symptoms to those whom circumstances compel to largely subsist on them. And yet it is a fact of experience that most of the fresh meats are much easier of digestion than vegetables, as they are more palatable and nutritious. Those who adopt vegetarianism, are usually the very individuals who are least able to bear it—whose temperament and manner of life demand as generous a variety of meats as they can obtain, whose only hope for health and usefulness and long life depend on a liberal supply of nutriment. There are men who can bear vegetarianism, and for a time, even semi-starvation : but, such persons rarely adopt any such rigid system of dietetics, unless forced to do so by hard poverty. There are many rough, wiry muscle-workers who, since they do only beast's work, can well thrive on beast's food—whose brains, like so much fallow ground, are simply an incumbrance, and who therefore chiefly need that kind of food which supplies the waste of the muscle. But brain workers are of a very different stamp, and those who for conscience sake adopt vegetarianism are usually as sensitive, delicate, and as intellectual as they are unfortunate and misguided. But in the light of science and experience, nothing is more absurd than vegetarianism for the feeble and dyspeptic.

I once dined in company with a young collegian, who, out of respect for the host, so far compromised his regulations as to allow himself to be helped to a dish of hash. From this he cautiously separated the potatoes, eschewing entirely the particles of meat, which were left on his plate. During the meal he entertained us with accounts of his wonderful feats at pedestrianism in his vacations. He could walk twenty miles a day without fatigue, and at that very time was on a trip to the White Mountains. As I looked at his pale and feminine features, tinged with an unnatural flush, (which I can never forget, though his name has passed from my memory,) I knew not whether most to admire the energy, the patience and the force of will that had enabled him to crucify his desires, or to pity the ignorance that to all human seeming must, if persisted in, destroy his usefulness. One year from that time he was in his grave, a victim to consumption. The lamp had gone out because there was no more oil in it.

I am not drawing an ideal picture. That unfortunate young man represents a not unimportant class of to-day. These lines may be read by some who, with all their might, are zealously striving to follow in his steps. It is to save such from an untimely death, or more likely from crippled usefulness, that I insist so strongly on the duty of living generously; for as I have stated, dietarians are not unfrequently the choicest and ablest. The sensitiveness of organization which is the basis of genius is frequently the source of ill health that impels to wild extremes. It happens in hygiene as in religion that the bold and original wander farther from the ways of truth than the stupid and indolent. Although in the majority of instances they return to the more excellent way, yet there are some who persevere to the bitter end, which is sure to come. With the strongest constitution it is a question of time how long the machine can be worked and wasted without nutriment to supply the waste, how long the fire can burn without fuel.

There are to be found those who might, if they chose, adopt Vegetarianism or Grahamism without serious detriment, who think so little and do so little with brain or muscle that there is little waste of tissue and consequently but little to supply. There are these who indulge so extravagantly in negative food—in the form of tobacco, or alcoholic liquors, or coffee, that there is very little desire or need for positive nutriment. But such persons are the last to adopt conscientiously, or perseveringly, any rigid system of dietetics. Consistent dietarians are usually men of susceptible temperaments and firm will. In this respect, their very fineness of organization becomes a misfortune. They are slain through their own energy and die by their own perseverance. If they were far less gifted, they would be more useful and more happy.

SUPERSTITION A CAUSE OF VEGETARIANISM.

Another cause for the popularity of Vegetarianism, as for many other exclusive systems of diet, was the superstition of the people.

The civilized and semi-civilized Americans abstained from meat for the same reason that the savage Africans abstain from certain articles of flesh or fish, because they supposed that religion forbade it.

The influence of enlightened Christianity has been to reform some of the worst of these superstitions among those by whom it has been received.

But still it must be allowed that the number of consistent dietarians is comparatively small. There are many who are dietarians in theory, but liberal feeders in practice. They suppose or maintain that it is a duty to deny oneself of all luxuries at the table, but practically they take the best that they can get. Sometimes they inwardly resolve to be true to their convictions, and abandon flesh and white bread, and all the other comforts of the table, but nature proves too strong, appetite asserts and regains its mastery, and they rise from every meal with a wounded conscience, until, in the course of years, by long unheeding, it becomes seared and silent. Consistent dietarians have this advantage over temporizers or backsliders, that with all their physical weakness and suffering, and impoverishment, they yet, so long as they live true to their theories, have the approval of the voice of conscience, which to them is as sweet and abiding a consolation as though it were based on truth and intelligence.

DRINKING AT MEALS.

In the use of liquid as of solid food, desire is the best guide. We should drink when we are thirsty ; and as we are usually thirsty at meals, especially when our food contains little water, we should drink with freedom, and usually to the full extent of the desire.

The dogma that has been laid down that we should

suffer from thirst while we eat, and drink only at the conclusion of the meal, because some of the lower animals do, is worthy of a system of medicine that sought to make a hell of earth by denying a drop of water to patients burning with fever, or hastened to the grave those dying for want of blood by taking out the little that remained ; or of a theology which made it a crime, under any circumstances, to be happy.

The quantity of food that is needed by any individual, climate being excluded from consideration, depends on these general conditions :

1. The organization.
2. The kind and amount of work done.
3. The quality of the food eaten.

Some men, and some races, with the same climate, and amid the same conditions, need more food than others. Large people do not necessarily eat and drink more than small people, for other conditions will vary ; but in the average they will certainly need more.

Small, thin people, sometimes, are very active and laborious, and so consume a quantity of nutriment that amazes those who think that size and a great appetite go together. In small, nervous organizations the force supplied by the food is expended in labor of the brain or muscle instead of being stored up. The thin man and the fat man may both have the same income of food, but the one spends as he goes, and the other is economical and accumulates.

Of the difference in the quality of food enough has already been said to make it evident that those who subsist on potatoes and cabbages need more in bulk of food—other conditions of organization, activity and

climate being the same—than those who subsist on beef,
mutton, fish, and grain.

Pereira estimates that one pound of beef is equivalent
to ten and one-quarter pounds of potatoes ; and it has
been observed that those tribes that subsist exclusively
or largely on vegetables and plants develop larger and
more prominent abdomens.

The story is told of a certain well known lecturer on
vegetarianism, that at a country hotel where he stopped
he disposed of nineteen potatoes at his breakfast.

The quantity that can be eaten by savages in cold
countries, after long fasting, is something stupen-
dous.

Bush gives the following account of the eating powers
of the Tungusians of Siberia :

"After pitching the tent and arranging the camp,
Telefont and Alexai, the two remaining Tungusians, sat
down to a *gallon kettle of hot tea*, and did not leave it
until they had emptied it of the last drop. They then
cooked a *four quart pailful* of boiled fish and soup, the
contents of which they also devoured. By this time
Zakhar, the other Tungusian, came up cold and hungry.
The same pail was then cooked *twice full* of boiled beef,
which the three emptied both times, even cracking the
bones for the marrow. Then, after rinsing the pail,
they cooked it full of 'crupa,' a kind of mush, which
they ate as soon as it was prepared. After all this,
either their appetites were not fully appeased or else
they feared to break off too abruptly, for they commenced eating dried whale, even devouring the fish
skins, which they first broiled in the flames of the
fire. After this they went outside of the tent and

began the preparation of other food, and the last thing heard after retiring was the cracking of beef bones for the marrow."

Stories equally, if not more, amazing are told by other travelers. Captain Cochrane says that five Ya kuti of Siberia will eat a calf weighing two hundred pounds at a single meal. Ross estimates that the Esquimaux eat daily twenty pounds of flesh and oil.

The *second* law, that of variety—is as powerfully enforced by experience as the preceding.

Variety stimulates the appetite, and with it the organs of assimilation. More food can be eaten, and with greater relish, when there are many than when there are only a few dishes. This fact, which all admit, has been presented against the view that there should be a variety of articles at each meal. Rightly considered, it is the strongest argument in its favor ; for the dangers and the temptations of overeating are nowadays far less than the dangers and temptations of undereating, and whatever induces us to eat liberally should be encouraged.

It is a fact of observation that meals composed of a number of palatable varieties, not only relish but digest sooner and with more satisfaction than a full meal composed of one or two varieties. For lunches and light repasts it matters little whether the variety be small or great.

The *third* law, that the food should be modified by race, climate, season, age, sex, occupation, activity, and the progress of civilization, is one that will, in general, be conceded without argument, and has already been demonstrated by the facts previously presented.

The general principle involved in the law follows from the explanation I have already given of the method by which mankind have found out what to eat and drink.

DIET MODIFIED BY RACE.

Some races, like the Germano-Scandinavians, are large eaters in whatever climate they may go. The Norwegians, the Germans, the English, Irish, Scotch, and the Americans, to a less degree, are devoted to the table. On the other hand, the French, the Italians and the Spaniards are comparatively moderate eaters.

It may be said in general that races that are hard drinkers are also heavy eaters.*

DIET MODIFIED BY THE SEASON AND CLIMATE.

As the diet of the tropics and poles is different from that of the temperate zones, so the diet of summer should be different from that of winter.

In the winter, therefore, we take more of meat and fats ; in the summer more of fruits and cereals. The appetite guides rightly in this as in all other matters of diet. A greater quantity of food is required for winter than for summer, for the very clear reason that more heat must be evolved in order to keep up the temperature of the body. The desire for food often increases with the increasing cold of autumn and diminishes with the diminishing cold of spring.

---

* See my work on " *Stimulants and Narcotics* " of this series.

DIET MODIFIED BY AGE.

That the young and growing need more food in proportion to their weight than the matured and aged, is one of the unquestioned facts of observation ; but there are very few who realize how old we are before we stop growing.

We do not reach our full height and breadth until twenty-five or thirty, and in some instances even later. The period of most rapid growth is in the early years, and during this period, and especially that portion of it which is called boyhood and girlhood, the amount of nutriment that can be taken is very great. Children need to take food more frequently than adults, and especial care should be taken that they do not long go hungry, for meager or insufficient food in the days of growth results in stuntedness, or dwarfiness, or disease that no subsequent care can fully remove.

The popular theories that children should not eat what they crave, that they should live in a state of subdued starvation, that they should eat only one or two varieties at a meal, that they should avoid what they like best and confine themselves to what is more disagreeable ; that they should not drink with their meals—these and other similar unscientific and more pernicious heresies took their origin in the same soil from which Grahamism arose—namely the reaction from the gluttony and sensuality of our Saxon ancestors.

Children need guidance in the selection of their food, but they should be guided by wisdom and not by ignorance.

The importance of the diet of children no words can

food depend their future size, health, and strength of
muscle and brain.

Children bear over-feeding much better than underfeeding.   The evils of repletion are evanescent ; the
evils of starvation are permanent.   Repletion carries
with it its antidote ; starvation has no cure except in
food.

The children who gorge themselves with ripe and
mellow fruits are less injured than they or their friends
suppose ; and far less than those who are kept half
starved.   The most distressing fact connected with the
extreme poverty in our great cities is the meager nutriment of the children, as it is revealed by their wan faces,
and their stunted forms.

Everywhere, in all countries, the poorer classes are
smaller, thinner, and weaker than the rich ; and one
chief reason why they are thus more favored is that
they are more liberally nourished, especially during
the period of growth.

For infants, as all know and admit, the best food is
milk ; and to this, according to growth and strength,
solid food should gradually be added.

It should furthermore be remembered, that our constitutions change so radically during life, that articles
of food which at one time are injurious, become most
agreeable and beneficial.

DIET MODIFIED BY SEX.

Women are smaller and lighter than men, and therefore, other conditions being the same, would need less
quantity of nutriment.   Other conditions are, however,
not the same, for in some countries woman works more,

and in other countries less than man. In barbarous lands woman is the slave of man and performs menial tasks ; in civilizations a little advanced, she often shares with him the labors of the field, shop, and the counting room ; in the most enlightened nations woman is the toy, the companion or ornamental appendage to man, and in certain social states, uses her brain but little and in trivial matters, and her muscles scarcely at all. The brain of woman is about one tenth less in weight than that of man ; and the amount of brain-work of the more severe kind, is incomparably less than that which man performs.*

As I shall presently show, those who toil hard with the brain need a liberal supply of first-class nourish-ment. In proportion as woman thinks less than man, in that proportion, so long as she uses her muscles but little, she will need less food than man.

Observation shows that woman almost everywhere, in civilized lands at least, eats less than man ; and in countries like the United States, where she is nervous, and sickly, and slender, she needs a different *quality* of food from that which is agreeable to the hardier sex. The wives and daughters of farmers, who during the winter perhaps never leave the house, and who, forgetting the example of their mothers before them, with strange perversity prefer the parlor to the kitchen, the piano to the cooking stove and wash tub, must not only eat less, but eat differently from their husbands and brothers who toil from sun to sun in the open field.

* According to calculations that I have made, based on the biography of the most distinguished men and women of all ages and nations, about *one fiftieth* of the best work of the world has been done by woman.

FEMALE BOARDING SCHOOLS.

The diet of female boarding schools is of sufficier importance to entitle it to separate and special. consic eration, and a volume might well be devoted to it.

Among the manifold causes of the delicacy and nerv ousness of American women, slow starvation at schoc is one of the most prominent. In obedience to the old but fortunately waning superstitions, that the mind ca be cultivated only at the injurious expense of the body that whatever is pleasant must necessarily be pernicious and that the benefits of any system of diet, as of exer cise, is exactly in proportion to its disagreeableness, th managers of boarding schools have prescribed diets ries that oftentimes more than neutralize the goo effects of their teaching.

Fashion has joined hands with superstition, an through fear of looking gross or healthy, or of incurrin the horror of the disciples of Lord Byron, our youn ladies live all their growing girlhood in semi-starvation they become thin and poor, their nerves become pain fully sensitive, and when they marry they give birth t starvelings.

DIET MODIFIED BY INDIVIDUAL TEMPERAMENT.

All the other conditions of race, climate, season, ag sex, being the same, or as nearly the same as possible the food varies with the temperament, in a manne at once striking and mysterious.

In the chapter devoted to the description of the lead ing articles of food, some remarks were made on the palatability, digestibility and nutritive value. Thes remarks were of necessity general in their character ; to all these are individual exceptions.   There are idiosyncrasies in the matter of diet that defy explanation, and probably can be understood only when the mystery of life itself is solved.   This applies not only to positive but to negative food, and to all our principal medicines.

Even quinine has been known in a few instances to produce a peculiar and disagreeable eruption on the skin.   Some cannot bear mutton, others are made ill by a pear, or watermelon, or cucumber.

All these peculiarities are strange enough, but no more strange than other peculiarities of appetite or taste.   Why it is that one likes tomatoes, or peaches, or ice cream, or liver, or melons, or brown-bread, and another is indifferent to all these things, or is, perhaps, disgusted at the sight of them, is a problem as unsolvable as the origin of existence.

To those who are impatient that such caprices are not explained, the best reply is, to say that they are no more mysterious than that we should exist at all.

One thing is clear, that they must, to a certain extent, be abandoned ; and those articles that poison any of us must be refused, even though they be food to all others.

DIET MODIFIED BY HABIT.

As by long practice one who at first is repelled by the odor of tobacco, can become so accustomed to it that ten pipes or cigars are a pleasure and a luxury, so many varieties of ordinary nutriment can be forced on the system, and the system can be gradually adapted to it, so that what originally was pernicious to it becomes nutritive and agreeable.   The truth of this statement is demonstrated by daily experience of individuals, and by the experience of humanity in all parts of the globe. Travelers are witnesses to the truth of this statement, for they wear themselves to the dietetic habits of the people among whom they reside, and are frequently benefited rather than injured by these radical changes of life.

It is a part of the conditions of life ; a part of the law of evolution, that we should be able to adapt ourselves to the diet that we find, as much as to all other conditions of environment.

In Africa some become so accustomed to the terrible *mboundou* poison * that they can drink it by the bowlful without harm.

DIET MODIFIED BY THE PROGRESS OF CIVILIZATION.

It is an essential inference from all the facts that have been advanced relating to the eating customs of different countries, that the food of civilization must be different in quantity and quality from that of barbarism.

The thoughtful, cultured European or American is a being as different from the savage African or Australian, as the Africans or Australians are different from the higher order of apes ; and correspondingly his food must differ from that of the savage, as the food of the savage differs from that of the ape. Nay, more, the gulf that separates Shakespeare and Newton from the

---

* This drink is prepared by scraping the *mboundou* root in a vessel, and pouring in about a pint of water ; fermenting at once takes place, and the liquid effervesces like champagne.

Papuan is wider than that which separates the Papuan from the gorilla and the chmipanzœ ; and therefore it is easier for the lowest order of human beings to live after the manner of the apes than for the highest orders of humanity to live after the manner of the savages.

The difference between the savage and the civilized consists mainly in the larger, richer, and finer development of the brain and nervous system.

The advantage of the enlightened European or American over the lowest races in size of brain is great ; but the advantage in quality of brain is far greater. This finer quality of brain, in highly advanced races, is revealed by the correlated conformation of the features, by the fineness and softness of the skin and hair, and by general sensitiveness.

The trustworthy descriptions that travelers give us of the insensibility and coarseness of savage tribes are amazing, and show clearly enough that all civilization is purchased at the price of pain and sorrow ; that all refinement has its compensations.

When we read the accounts of the brutality of savages to each other we shudder, and perhaps lay aside the book with nausea and faintness. Among many African tribes the feeling of love between man and wife, as we understand it, is not known ; and it is as customary for husbands to flog their wives, as among us it is to kiss them. Whips, made of hippopotamus hide, are made for this special purpose ; and if a husband neglects to flog his wife, her relatives complain that she is badly treated. This custom prevails among a large number of barbarous and semi-barbarous races, and in different parts of the world ; and the custom of flogging or bambooing, and otherwise pounding the body for slight offences, is common to all except the best cultured nations.

In assuming, as most of us do, that these apparently cruel processes inflict anything like the amount of pain on the coarse and brutal people on whom they are usually employed, we make a great mistake. They probably suffer less from these hideous tortures than we do from reading the account of them.

Parkyns says that the coolness with which the Abyssinians receive the punishments inflicted on them by their Turkish governors is wonderful ; and he rightly attributes this coolness not to mental endurance, but to *physical incapacity to be pained.* Hundreds of blows even aged culprits will receive without even crying out. For a trifling reward they would be willing to take five hundred blows with the lash. Still more markedly that their insensibility is proved by the voluntary treatment they give themselves or each other. Their duels, which are "engaged in by young men on the slightest possible pretext," are conducted with a hippopotamus hide, which makes furrows in the skin, and draws blood with every stroke. An Abyssinian belle gashes her body in order to raise beautiful scars, which are there considered ornamental. For bracelets they tie a corrosive root around the wrist, which eats into the flesh and raises a perfect band as thick as one's finger ; and probably all this causes less pain to her than the mere pressure of an artificial bracelet on the growing arm of an American maiden.

The stolidity of the North American Indian under tortures that are apparently cruel, and the coolness with which the Hindoo swings by hooks in the flesh, rolls on the ground for long journeys and over rough roads, and commits various and horrid acts of religious devotion, are quite familiar, and may all be reasonably explained. The tattooing so commonly practiced among wild tribes, and the disgusting rites of circumcision, which for immemorial ages have been practiced among African tribes on little children of both sexes, probably cause little if any more pain than they would on the monkeys among whom they live, and to whom they are believed to be so closely related. The African negro, when he wishes to break a stick, breaks it over his own head instead of on his knee, as is the custom with us.

Then again, among many wild people, child-bearing, which makes a modern woman an invalid for days, and weeks, and months, is usually all over in from half to three quarters of an hour, and is not attended with any cries or tears, and the mother right away resumes her menial duties.

To a less degree the same holds good of other nations. It is clear enough, also, that the ascetic customs of the early Christian and mediæval ages ; the protracted floggings of our navies ; the barbarous tortures of the Inquisition ; and the hideous civil punishments of various degrees of past centuries, caused immeasurably less actual pain than they would if inflicted on the present descendants of those sufferers.

We have only to go back one hundred, or one hundred and fifty years, to find in England lunatics treated as criminals and punished with whips, and blows, and chains ; women publicly flogged at the whipping post, branded with hot irons, and driven at the cart tail ; thumbs tied with whip cord ; both sexes and tender ages hung for various offences so frequently and so publicly that the London of the eighteenth century has been rightly termed "the city of the gallows!" duels occurring continually and for the slightest pretexts; men and women taking part in public sports with the cudgel and broadsword, bear and bull baiting, boxing and cockfighting.

In nearly all barbarous nations, woman is the slave of man, and performs the most toilsome and menial tasks, but in this position her positive sufferings have probably been much overestimated. The modern wife not unlikely experiences keener distress, to her at least, from anxiety about the servants who pretend to do her work.

Even very recently a great change in methods of punishment have been inaugurated. A schoolmaster who should flog the children of the upper circles of New York, as was everywhere done in this country twenty-five years ago, would be dismissed the first week.*

The bearing of all these facts on the question of diet is sufficiently apparent.

In the evolution of humanity all departments progress together and in harmony with each other, and no one can become refined much in advance of the rest. Diet, like legislation, like social amusement, like religion, must change as the race advances in the direction of refinement.

The organization of the average European and Amer-

---

* The story is told of a certain teacher in the Edinburgh High School during the last century, that he was accustomed to call up twelve dunces, place them in a row, and strike each one on the back in succession, so as to make him cry out. The variety of musical notes thus produced by what he called an "organ," was so pleasing that other teachers were invited to the entertainment, and they returned the compliment on their own pupils.

ican of the cultured classes of the present day is so much finer and so much more sensitive than it was one century, or even one quarter of a century ago, that they cannot bear the same medicine, the same stimulants or narcotics, or the same ordinary food that they could then. What was then appropriate and necessary and right for their coarser organizations, is, for the finer organization, that has developed with the intense brain-work of our time, suicidal or cruel.

If our better classes should persist in using alcohol or tobacco, or pork, in the manner of their ancestors, they would soon become exterminated.

It is astonishing how savage and semi-savage races, and coarse, thick-skinned organizations everywhere, can bear those stimulants and those articles of positive nutriment which, by the ruling classes of to-day, and especially of this country, must be used so cautiously. So far as I can estimate from the many writers of travel that I have consulted, there are over six hundred millions of people—half the population of the globe—who smoke tobacco or opium, or both, through nearly *all their waking hours*. The *four hundred millions* of China—men, women and children—smoke almost as constantly as they breathe ; even during the meals, if they stop for a moment, and in the night if they chance to wake, they seize a pipe and take a few whiffs. They cease to smoke only when they cease to live, and when the sick man no longer asks for his pipe, the attendants prepare for his funeral.

Burton, speaking of the East African, says :

" He drinks till he can no longer stand, lies down to sleep, and awakens to drink again. Drinking bouts are solemn things, to which the most important business must yield precedence. They celebrate with beer every event—the traveler's return, the birth of a child, and the death of an elephant—a laborer will not work unless beer is provided for him. A guest is received with a gourdful of beer, and among some tribes it is buried with their princes."

"The highest orders rejoice in drink, and pride themselves upon powers of imbibing ; the proper diet for a king is much beer and a little meat. If a Mnyamwezi be asked, after eating, whether he is hungry, he will reply Yes, meaning that he is not drunk. Intoxication excuses crime in these lands."

Livingstone says that at Angola funerals constitute one of the principal recreations, and they are attended with debauchery, feasting and intemperance. If on these occasions a native is reproved for being drunk, he will reply, "Why, my mother is dead," as though no other apology were needed.

No longer ago than 1805 it was said of the inhabitants of certain districts of Scotland that they experienced delight on hearing of the death of a man or woman because of the prospect it afforded them of getting their fill of whiskey ; and men died, saying "they would not be happy unless men were drunk and fought at their funerals."

Lewes says, in 1800 "it was no unusual thing to be a 'three bottle man' in England or Germany."

In Austria, at the beginning of this century, it was said that the dinners would sometimes last four or five hours.

A few centuries ago our Anglo-Norman ancestors

began every important enterprise with banquets, many
of which were riotous and drunken orgies ; and of the
Icelanders it is said that in their feasts they indulged·in
the most unseemly exhibitions, and ended with throw-
ing the bones at each other across the room.

The ancient Grecians, according to Tacitus, never
undertook any great affair without a feast. In modern times the remains of this custom are preserved
in the habits of business men, who discuss their business schemes in France at breakfast, and in England at
dinner, and in the well-known banquets of philanthro-
pic organizations.

The Americans of the present day, of both sexes, bear
less of alcohol and less of tobacco than any other people
on the face of the globe. There are tribes in Africa who
drink their "*pombe*," or plantain wine, from early dawn
to bedtime, with a perseverance that puts the beer-drink-
ing Germans far in the shade. But these excesses in tobacco and alcohol rarely seem to bring on diseases of the
mind or nervous system, such as a very small percentage of indulgence in these articles does with us. Among
all these coarse races, whatever their habits of eating and
drinking may be, insanity and paralysis are exceedingly
rare affections.

The same difference of susceptibility to stimulants in
kind, though not in degree, is seen in contrasting the
present generation of Americans with their immediate
ancestors. Few are the matrons of our time, among
the well-to-do orders, who could smoke their pipes, or
take their snuff, without immediate and serious harm ;
and there are few men among the same classes who
could take their daily and hourly drams in the manner

of their fathers.* The notion that the greater susceptibility of the present generation to alcoholic liquors is
due to the fact of adulteration, is mostly untrue, for, as I
have elsewhere shown, most of the adulterations of
wines and liquors are comparatively harmless; and
whatever of harm comes from their use in these days
must be charged to the alcohol they contain.

THE TREATMENT OF DISEASES HAS CHANGED.

Diseases also follow the great law of evolution, and
as humanity moves onward, some of the more virulent
and fatal inflammations and affections grow milder and
milder, and perhaps entirely disappear, while others,
especially of the nervous variety, increase and multiply.
The changes in the treatment of disease follow closely on the changes of diseases themselves. That the
frequent and abundant bloodlettings, and purgings, and
the starvation diet of the last generation were always
harmful, no careful student of history or of medicine will
claim; rather is it to be believed that they were on the average as successful as the average of human skill could
expect to be; but the physician who should now systematically treat his patients by these methods, would be tried
for manslaughter. Where formerly we reduced, we now
sustain; where we once abstracted blood, we now trans-
fuse it; where formerly we prescribed purgatives and
fasting, we now prescribe tonics and feasting; and in
thus changing our practice with the changes in disease
we do wisely and well.

---

* In the reign of the Tudors the leading merchants transacted their business over their cups in taverns, as is now done among the lower classes.

DIET OF THE PRESENT DAY.

Now what is true of punishments, of sports, of stimulants and narcotics, is just as true of ordinary food ; we cannot and do not bear the coarse and ill cooked diet of the last generation. The old Northmen ate raw flesh, and it is believed that the ancient Gauls were cannibals. Just as claret, Rhine wine, and cigars, and cigarettes have taken the place of port and brandy, of long pipes and strong tobacco, of chewing and snuff-taking, so the oatmeal porridge, the rye and Indian bread, the salt pork and beef, the eels and smoked fish on which our fathers thrived, have given place to white flour bread, light biscuit, fresh and tender beef and mutton, and delicious fruits—all prepared and cooked in a style that harmonizes with our capricious appetites and susceptible organizations.

The remains of the old customs of eating and drinking are yet seen among the distant rural classes, and. among the abjectly poor of the cities, especially among our emigrant population. The different nations vary much in their susceptibility in this respect. The sour bread, cold sausages, strong cheese and abundant lager beer on which our German friends thrive, would drive the average American to despair.

To what extent this refinement of organization and corresponding refinement of dietetics and cookery will go, no one can now well foresee.

The *fourth* law requires that our meals should be *leisurely* enjoyed, at a pleasant *social* table.

All writers on hygiene—even the very worst—agree that it is better to eat slowly than hastily ; that the food should be well masticated, and leisurely swallowed. This is one of the very few directions found in our popular treatises on diet, that are in accord with recent science, and sustained by experience. There is no doubt that rapid eating—and here I heartily unite with those writers whose views or theories on almost on all other subjects of hygiene I have been obliged to combat—is one of the greatest evils of the American social system. It is the result partly of our climate, partly of our institutions, and partly of the strifes and the necessities of our pioneer life. On the continent of Europe, as all travelers know, the habits of the people are very different; even the peasantry of France and Germany take more time at their meals than the laboring classes with us. National differences are also observed in the habits of taking beverages. The custom of drinking, or rather guzzling at public bars, is not recognized among the better orders of foreign society, and in England is allowed only to the poor and degraded. When the Parisian desires a glass of brandy, or cordial, he sits down in a *café* and takes is time for the luxury; meantime reading a newspaper or chatting with a friend. The Yankee would gulp down twice the quantity of liquor, while the moderate Frenchman was giving his order to the waiter. When an American soda-water establishment was first opened on the *boulevards* in Paris, crowds of Frenchmen used to stand before the doors and windows, laughing and jeering at the novel sight of men pouring down a tumblerful of liquor at a swallow. It was a long time before the French people could be induced to partake of the very agreeable luxury, and as soon as they began to patronize the establishment in large numbers, it was found necessary to provide tables

and chairs, where they might sit and sip the foaming
soda-water as they were accustomed to sip their cordials
and *liqueurs.**

If the habit of pouring down fluid, even pure water,
into the stomach, is injurious, how much more so must
be the shoveling in of solid and half-masticated food.
Rapid feeding overtaxes the stomach and interferes with
digestion, just as a too rapid flow of ideas upon the
mind overtaxes the brain and interferes with the successful performance of its functions.

To preach on this subject is almost idle. A man who
dines at a table where all his companions eat on the
jump, finds it as hard—and I may say impossible—to
act independently as it is to walk slowly amid the press
of lower Broadway, or to saunter at ease in the ranks
of a flying army. Haste is as contagious as some of our
worst diseases, and very few can suffer long exposure to
it without showing the effects of the poison.

The only hope of a reform in the dietetic habits of the
Americans lies in the gradual development of a better
fashion of dining that now prevails among us. The difference between the habits of the English and Americans in
this respect, is as wide as the Atlantic. An English
dinner, to the American who partakes of it for the first
time, is an event in his history. The courses are so arranged and subdivided that it is impossible to eat more
than a little at a time, and at the end of the meal—
which usually occupies from one to two hours—he
hardly knows whether he has been eating at all. The
time passes agreeably in social converse, and the stomach
receives its burdens so gradually and imperceptibly that

* See my work on *Stimulants and Narcotics,* where the relation of American
climate to American nervousness is fully discussed.

it has full time to marshal its forces as they are needed, and thus digest with ease an amount of food which, if hastily swallowed, would cause the direst distress.

Savages, who eat with their fingers, are always greedy and rapid eaters. The introduction of forks in the 17th century has contributed much towards a more calm and refined habit of dining.

EATING CLUBS.

The poorest club conceivable is better than solitude. Those whom circumstances force to board themselves, do well to keep each other company. Two or three, or even more, can join their forces, thus providing larger variety, greater comfort, better appetites, and more prosperous digestion. For many organizations, solitary dining is slow death. There are those who can bear it, just as there are those who can bear vegetarianism, abstinence from muscular exercise, or confinement in impure air ; but the suggestions here given must be for the average, and not for the exceptions. And yet we should bear in mind that those who survive long-continued violations of hygienic laws, might perhaps have been even more sturdy, and achieved even larger success, had they lived more in conformity with those laws.

RESTAURANTS.

A disadvantage of restaurants is that they compel their patrons to select dishes by the names on the *carte*, and not by their appearance and flavor when brought upon the table.

The great objection to dining by a bill of fare is, that we cannot tell what we most desire until we see the

articles of food, and inhale their savory fragrance.
French names, with high prices annexed, are at best
poor appetizers. This is the philosophical explanation
of the fact that, in sitting down to a public table, we
often study over the schedule in nervous despair, and
then decide upon a dish, which, as soon as it is placed
before us, we find we have no relish for. Nothing can
redeem the life at a public table but a pleasant circle
of very dear friends, to share our meal with us, as is
the custom in Paris. The maxim, "Chatted food is ill-
digested," needs only the substitution of one word—
well for *ill*—to make it thoroughly true.

The three best digesters are, sound health, a good
table, and pleasant conversation; but the greatest of
these is conversation, for it can divert the mind even
when the health and food are both unsatisfactory. In
the charming biography of Charlotte Brontë, by Mrs.
Gaskell, we are told that the father of the accomplished
authoress, on account of a weakness of digestion, was
accustomed to take his meals by himself, apart from the
rest of the family. If that had been his habit long, it
is no wonder that he was often compelled to give vent
to his attacks of hypochondriasis by "firing pistols out
of the back door."

TIMES OF EATING.

The frequency with which food should be taken depends on various conditions. It varies with different
races, and in different climates; and according to the
amount of labor performed, and the quality of the
food.

This, as well as other questions of diet, must be
answered by *experience*.

The custom of taking three meals daily, which is so generally observed in the United States, is, on the whole, a wise one, since it has taken its origin in the accumulated experience of many years.

It is not well, however, to observe this custom too arbitrarily, for a fourth meal is oftentimes a positive advantage, even in our climate, where less food is required than in Northern Europe.

The Germans eat four and five times daily, and in England very many take four meals as regularly as we take three, the fourth consisting of a light supper, on thin slices of bread and butter and a small cup of tea, and among the country people of substantial pastry and meat.

This fourth meal is taken between nine and ten o'clock.

Since among the middle and upper classes of England, the late dinner, which begins at 7 or 8, is prolonged to 9 o'clock, the supper and dinner come very closely together. In the United States a fourth meal is taken much more frequently than is believed, both in city and country ; but unfortunately the materials of our evening entertainments are too often both unpalatable and indigestible.

On the ocean steamers four or five meals are the order of the day, and except by the sea-sick, are well patronized, without discomfort or injury.

I should add that in rural districts a lunch in the middle of the forenoon is oftentimes expected, and in the haying season milk is drunk instead of water. Our customs are not so utterly different from those of Europe as might be thought. While the English eat so largely, the French, who are separated by only a narrow channel, are very moderate at the table, and do not take more than two full meals daily. In the morning, on rising, a cup of coffee with a very little bread satisfies them until about 11 o'clock, when they take breakfast, and at 6 o'clock comes their well cooked and delicately served, but not over abundant dinner.

TIMES OF MEALS HAVE CHANGED WITH THE PROGRESS OF CIVILIZATION.

The custom of having regular hours for meals, is peculiar to civilization. It is only possible where food is tolerably abundant and accessible, and it is most necessary among those who are most sensitive, delicate and nervous.

The savage can go for days without anything to eat, and then can, without injury, gorge himself like a boaconstrictor.

The old Highlanders originally had but one meal, and afterwards two meals daily.

In England several centuries ago, among the "great families," there were four meals daily : breakfast at seven, dinner at ten, supper at four, and "livery," corresponding to modern "tea," at eight or nine o'clock. They sat at dinner three hours, from ten till one in the afternoon.

In Thibet there are no regular meals. The family never assemble, but "eat when they are hungry, drink when they are dry." Like the North American Indians they eat out of a huge pot in which the meal is boiled, and each one snatching what he wants.

The Kisawahili has no words to express breakfast, dinner, and supper.

Savages seem to care much less for the gratification of the taste of food, than for the appeasing of the hunger. With them the pleasure of eating is of a negative rather than of a positive kind ; hence they eat rapidly, voraciously, in order to fill up the stomach as soon as possible. The relics of this habit can be seen readily enough in any cheap eating-house of New York.

Eating for the refined gratification of the taste *per se,* is peculiar to high culture.

The civilized man who undertakes to imitate the savage in these alternations of fasting and feasting, would soon bring his wretched existence to an end ; for no fact of hygiene is better recognized than that after very long abstinence, food *must be taken slowly* and *in small quantities.* The longer and the harder we have worked on an empty stomach, the greater the necessity of caution at the next meal. Just as it is possible for the brain to get so tired that we cannot sleep, just so it is possible for the stomach to get so tired that it cannot digest.

MEALS OF THE ROMANS.

The ancient Romans had but two meals daily—dinner and supper. The dinner, or *prandium,* was taken in a standing position, at nine o'clock, and usually consisted of the remnants of the supper of the previous day. The supper, or *cœna,* the great meal of the day, was taken about three or four o'clock in the afternoon. This consisted of three parts : a *gustus,* to sharpen the appetite, as raw oysters on the shell are used in our day ; the *caput cœnœ,* made of a variety of courses ; and the *mensa secunda,* or dessert, consisting of pastry and fruits. On these dinners enormous and fabulous sums were expended. Heliogabalus spent $20,000 for a single dish. Œlius Verus wasted $250,000 on a single entertainment to which only twelve guests were invited, or *over $20,000 for each individual.* It is stated that Vitellius spent over $16,000 daily for his supper.

The dinner became gradually later and later in the day, and then the morning meal, or *breakfast*, was introduced. King Canute, it is said, first established the custom of four regular meals daily.

THE BEST HOUR FOR DINING.

The oft asked question whether it is better to dine at the middle of the day or in the evening, must be answered by the occupation and circumstances of the individual.

It is of all importance that the principal meal of the day should be taken leisurely and without harassment; and if the hour of leisure that is required for a comfortable dinner cannot be found in the middle of the day, it is better to wait until later.

In this respect the present customs are just about what they should be. Business men in the large cities dine at five or six o'clock, for the two-fold reason that the middle of the day is the busiest hour, and because most of them are so far from home that they must take their meals in restaurants. In the olden time, when business was carried on quietly and on a small scale, and merchants lived over their stores, it was easy and natural to dine at twelve o'clock ; but under the new dispensation, the custom has been necessarily abandoned. In the country, where toil is easier and calmer, where even in the haying season a good round hour is

allowed for "nooning," and where they rise so early that
twelve o'clock is about as late as four or five is for the
residents of the city; where they go to bed with the
chickens, and where, furthermore, there is comparatively
little to harass the mind, in the morning, at noon-
day, or at night, it would be absurd, and to many injurious, to postpone the principal meal much beyond
midday.

MIDDAY LUNCHES.

There are very few civilized beings who can go from
morning till night with nothing to eat and not suffer.
Such prolonged fasting is the prerogative of savages.
There can be found now and then a business or professional man in whose arteries the blood of his
pagan ancestors is so little diluted that he only feels
hungry twice a day, and there are others who abstain
from midday lunches because they believe it to be a
virtue to go without eating. For the majority of
Americans this mistake is a most fearful one, and has
caused innumerable woes. Dyspepsia in all its phases,
nervous diseases of all kinds, and death itself, are the
rewards that nature is continually bestowing on those
who thus refuse her bounties.

Abstinence from regular meals in health is a vice
which only professed gluttons should indulge; but if
one must lose a meal let it be the last one of the day,
and not the breakfast or lunch. It is not always necessary that the lunches should be liberal or varied; for
many, a few mouthfuls suffice—or at least, stay the
stomach until evening; but, as a rule, it is well for
those who dine late in the day, to take *a substantial lunch,*

in which meat, cold or warm, or fish of some kind, or nourishing soup, as well as bread and butter, shall be represented.

The danger of spoiling the appetite for dinner is infinitely less than the danger of so weakening the stomach by long abstinence that a good dinner cannot be well assimilated. It is far better to overeat at lunch than not to eat at all.

EATING BETWEEN MEALS.

I am often asked whether it is well to eat between meals. I usually reply somewhat in this wise: To take something between meals, is better than to go long hungry—better than to overload the stomach at the table, and better than to waste precious hours in settling our doubts on the subject; but it is better still to make our ordinary meals so complete and so satisfactory that there will be neither desire nor necessity for taking anything substantial during the intervals. Those who are most tempted to indulge the appetite at other than the regular seasons are dietists and vegetarians who live in a kind of chronic semi-starvation, in which the need of the system is never fully met, and consequently the appetite never thoroughly appeased—who are ever hankering for something to eat, because the waste of tissue or nerve is not fully repaired. Such persons are never at peace. With them duty and disease are ever in perpetual guerrilla warfare, in which both sides by turns are victors. An intelligent, ambitious, industrious man, who daily enjoys three agreeable meals, needs no such warfare as this. It is wrong to suppose that we should always keep ourselves on a war

'ooting, in the matter of diet. Right habits should be
'ormed under the guidance of science and experience,
and calmly, patiently strengthened until they become
automatic.

EATING BEFORE RETIRING.

Again, I am often asked whether it is well to eat just
before retiring. My reply is very simple. It is better
o eat than to retire very hungry. Nothing is gained
y keeping the stomach so long empty that hunger
mounts to positive pain. Prolonged abstinence weak-
ns the whole system even more than gluttony and
intemperance. When our meals are what they should
e, and when we retire at the proper time, we do
ot usually need or desire anything substantial at
ed-time ; but when the supper is very light and the
uties of the evening prolonged and arduous, the mid-
ight hour will often find us in greater need of food
han at any other time of the day. To deny ourselves
nder such circumstances, and retire hungry because
, is not our accustomed hour for eating, is both cruel
nd injurious, for excessive hunger is a foe to sleep.

Some authors and public speakers are in the habit of
voiding the principal meal of the day until their even-
ig labors are over. I cannot find out that they are
ijured thereby. The disadvantage of taking a full
eal at so late an hour is more than compensated by
ie greater freedom from care that is enjoyed at that
our, for food taken with a heavy heart or overburdened
rain weighs painfully in the stomach and is digested
ι sorrow.

But the general rule should be to make an evening
eal—if we indulge it—of food that is readily digested,
ιch as oysters, or crackers, or bread and butter.

CHAPTER VI.

BRAIN-WORKERS—whether literary, professional, or business men—need the best of food, served in the most agreeable manner, and in variety and abundance, and for the following reasons :

(1.) Labor of the brain exhausts the system more than labor of the muscles. According to the estimates of Prof. Houghton, three hours of hard study produce more important changes of tissue than a whole day of muscular labor.

Whether this statement is mathematically accurate or not, we do certainly know by experience that a few hours of mental labor is more exhausting than a whole day of muscular labor to those who are accustomed to such toil. No literary man can spend as many hours at his work as the day laborer. While the mason, the carpenter and the haymaker work their ten hours a day, with only moderate fatigue, the professional man is wearied by three or four hours of severe, consecutive thought. This exhaustion that we feel after hard study is the result and concomitant of the waste of tissue. This waste of tissue is supplied by food.

If the theory of the correlation and conservation of

forces be carried to its logical conclusion, it would seem
that for every mental act there is a corresponding expenditure of force which bears a direct ratio to the
thought evolved, and that unless the proper force is
supplied, thought becomes impossible, just as steam is
impossible without heat, or motion without some force
to produce that motion.

Phosphorus, which is a prominent ingredient of the
brain, is deposited in the urine after mental labor, and
recent experiments have shown that by the chemical ex-
amination of these phosphates deposited, it is possible
to determine whether an individual has been chiefly
using his brain or his muscles.

That the brain is the organ of mind is now as well
established as any fact of science. The brain, being
the noblest organ of the body, receives a greater pro-
portional amount of blood than any other part, and is
of course correspondingly affected by the quantity and
quality of the nutrition. It has been estimated that
one fifth of the blood goes to the brain, though its
average weight is not more than fifty ounces, or about
one fortieth of the weight of the body.

2. Brain-workers as a class are more active than
mechanics or laborers. The literary man need never
be idle, for his thinking powers—the tools of his trade
—are always at hand. Bulwer, in his Caxtoniana,
mentions this fact as a great advantage that the literary
man has over all others. The mechanic has a definite
task, assigned for certain hours, and when that is over,
he feels free to rest. On the other hand, the powers of
thought and composition are only interrupted by sleep,
and the intensity of the labor is measured by our mental discipline and powers of endurance.

3. Brain-workers exercise more or less all the other organs of the body as well as the brain. Even the most secluded book-worm must use his muscles, to a greater or less extent, and the great majority of literary and professional men are forced to take systematic and vigorous exercise, in order to keep their brains in good working order. On the other hand, the uneducated and laboring classes, while they toil with their hands as their daily necessities require, are apt to let their brains lie idle, and thus the most important part of their nature undergoes comparatively little change, except that which comes from time and disuse.

KIND OF FOOD REQUIRED BY BRAIN-WORKERS.

If the brain could be used exclusively, without any exercise of the muscles, then the diet of brain-workers might be pretty exclusively confined to those articles which contain the fat, salts, and phosphorus, of which the brain is composed. But it is impossible to live by brain alone ; hence the necessity of a wide variety of food for the brain-working classes, of a quantity and quality adapted to nourish the *whole body*, with special reference to the nervous system.

The best food for the brain is fat and lean meat, eggs, and the cereals.

Is it now a matter of fact that brain-workers eat a better quality and larger quantity of food than mechanics and laborers. How is it with the different nationalities ?

We have seen that the ruling people of the world, who have from time to time shaped the destiny of humanity have always, so far as can be ascertained, been liberal feeders. This remark applies, of course, only to the ruling classes in these nationalities, and not to the slave or peasant class, who lived with them, but were not of them. But of the patrician or governing orders of society—the leaders of the world in legislation, in war, in commerce, in science and literature—it is pre-eminently true. The dominant classes among the Babylonians, the Persians, and especially the Romans, were free and luxurious in their habits of eating, although, in those days, there was less variety of food than at present. The Greeks, the most intellectual of ancient nations, were formidable eaters; and their repasts were greatly prolonged. Of the Romans it has been said, that no people were ever so devoted to the pleasures of the table. Among modern nations the greatest eaters are the English, the Germans, and the Americans—the ruling people of our civilization.— The diet of the Spaniards and Italians is notably less substantial than that of the English and Germans, just as their brains are less active and original.

Our standards, by which we measure nations and individuals, are too low and too narrow. I protest against the degrading spirit of materialism that would estimate a man by his weight on the scales, or by the number of years that it takes him to rust out. As if the human mind, with all its wondrous capacities, were created only to be imprisoned as long as possible, in a gross tabernacle of flesh, and nations were to be estimated, not by the thoughts they evolve, or the deeds of glory and usefulness they accomplish, but by the amount of adipose tissue their indolence enables them to hoard, or by the length of time that it takes them to die! Even the most ignorant hog-raisers studiously consider the *quality* as well as the quantity of pork that the different kinds of feed produce. And shall hygienists, in their estimate of the effects of diet on humanity, only look at the number of pounds avoirdupois that result from the different systems, or the number of years that the body can endure them? We argue that because the porters of the East, the native Hindoos, the Chinese, and the Irish peasantry, eat little or no meat, and are well, and muscular, and capable of a good measure of physical endurance ; therefore, all people, in all climates, and at all seasons of the year, should be vegetarians, and thus the world would be much better than it now is. The flaw in this reasoning is, that it takes too low and material a view of humanity, and ignores entirely the fact that, although the body can be sustained and kept from dissolution for a considerable period on simple fruits, cereals, and the like, yet in the history of the world nothing very great or good has ever been bequeathed to humanity by a nation of vegetarians.

GREAT THINKERS USUALLY LIBERAL EATERS.

With some exceptions the same facts will apply to individuals. The great majority of the leading thinkers and actors of the world—the philosophers, writers, orators, legislators, warriors, inventors, and creators of new eras in every department of human thought—have fed their brains with a greater or less abundance and variety of animal as well as vegetable food. We have, in biography and general observation, sufficient data from which to form a satisfactory and reliable opinion. Goethe was a vigorous performer at the table, and even to an active old age retained his fondness for good dishes. Says Lewes, his biographer :

"His appetite was immense. Even on the days when he complained of not being hungry, he ate much more than most men. Puddings, sweets and cakes were always welcome. He was fond of his wine and drank daily his two or three bottles."

On this diet, and amid great literary activity prolonged to extreme age, he lived to see his eighty-third year.

Of Peter the Great, Marmontel says that "he dined at eleven o'clock, and supped at eight ; an astonishing eater and drinker—two bottles of beer, the same quantity of wine, half a bottle, and sometimes a whole one, of brandy, at each of his two meals were scarcely sufficient for him, without reckoning the liquors and refreshments that he swallowed in the intervals."

I have studied with greater or less minuteness the biographies of nearly one thousand of the greatest thinkers of all nations and ages, and I find that aside from religious enthusiasts and some of the ancient philosophers who led very calm inactive lives, very few were known to be abstemious.

Dr. Johnson was a gormandizer, and when indulging in his favorite dishes this Great Mogul of literature displayed his eagerness by manifestations of satisfaction that we supposed to be peculiar to children and some of the lower animals.

When Charles Lamb was boarding, he sometimes invited friends to dine with him, paying the landlady a small sum. He observed that when Wordsworth dined with him, the landlady charged a sixpence more, and one day remonstrated with her on the injustice of such

discriminations, at the same adding that Wordsworth
was a great poet. "Don't know about the great poet,"
replied the landlady, "but I know he is a great eater."

The popular conception that those who think much,
eat little, is derived from the twofold fact that *some* of
the great philosophers have been comparatively abstemious, and from the fact that nearly all hard brain workers care little for their meals when they are in the midst
of severe tasks. In great crises they abstain perhaps
for several days, as did Elliott, the defender of Gibral-
tar, but pay day comes, and they must replenish their
wasted tissues or suffer the penalty. There is little
doubt that some authors have shortened their lives by
habitual underfeeding.

We can best arrive at the truth in this matter by
comparing different bodies or classes of men, and not
by selecting individual cases. Students in academies
and colleges, provided they are in good health, study
faithfully, and do not exhaust themselves by vices, eat
more, according to my observation, than young men of
similar ages, in ships, and behind the plow, and far
more heartily than mechanics and artisans. None
who board students, whether academical, collegiate, or
professional, ever regard them as light eaters. Those
exceptions who worry or fret themselves into nervous
debility, or who destroy themselves by vices, only prove
the rule.

Clergymen are also large eaters. Whatever their
theories may be, they practically acknowledge that
those who work with their brains need better nourishment than those who allow their intellects to lie idle.

I know very well that it is possible for some temperaments to study hard, for a limited season, on a

spare diet. There have been, and are now, hard students in our colleges, who, either from necessity or more likely from mistaken notions of hygiene, restrict themselves to a meager and unsatisfying allowance. With all such persons the evil result that must follow such a course, as surely as the night follows the day, is surely a question of time. There are those whose constitutions are so hardy, whose reserve powers are so abundant, that they can live, for a considerable time, on their capital. They can rise early and sit up late, and toil hard and long over their books, achieving the highest success in scholarship and literature, on an insufficient and unnutritious diet. But the pay day must come with them just as surely as with the poorest and feeblest, only it may be longer deferred.

To recapitulate in a few words : the diet of brain-workers should be of a large variety, delicately served, abundantly nutritious, of which fresh meat lean and fat, should be a prominent constituent. In vacations, or whenever it is desired to rest the brain, fish may, to a certain extent, take the place of meat. We should select those articles that are most agreeable to our individual tastes, and, so far as possible, we should take our meals amid pleasant social surroundings. In great crises that call for unusual exertion, we should rest the stomach, that for the time the brain may work the harder ; but the deficiency of nutrition ought always to be supplied in the first interval of repose.

CHAPTER VII.

In. athletic training the object is to reduce the fat, increase the size and hardness of the muscles, and the power of endurance.

Trainers have experimented with a variety of systems of diet in order to see what was best adapted to co-operate with severe muscular exercise, and regular habits to secure their ends. It has been found that the best bill of fare for athletes of all kinds—gymnasts, oarsmen, etc., during training, contains the following articles :

Lean and rare beef or mutton.
Stale flour bread.
Potatoes and other vegetables in moderate quantities.
Tea, coffee, and beer or wine in very moderate quantities.

This diet-table, it will be observed, contains little or no fat, and little starch or sugar, and therefore is not calculated to feed the fatty tissue. Its leading element is *nitrogen*, which is contained in the beef and mutton. Between beef and mutton there is little to choose. The quantity of tea and coffee allowed is always limited —usually not more than a single cup at each meal— and of the two, tea seems to be generally preferable to coffee. Of ale or wine, a glass or two are all that is allowed.

Weston, in his great walk, ate those article that best suited him, but used no alcohol.

The query may be raised whether the vertigo from which he suffered, and which caused his failure, might not have been prevented if he had taken some wine or beer, instead of so much tea and coffee.

King, an English athlete, in training, took for his *breakfast* two lean, rare mutton-chops, stale bread, and one cup of tea without sugar; for *dinner* one, or one and a quarter pounds of beef or mutton, toast or stale bread, a very little potato or other vegetables, half a pint of old ale, or a glass or two of sherry, or one cup of tea without sugar, or eggs and dry toast ; for *supper* half a pint of oatmeal porridge or half a pint of old ale.

The effect of this exclusive diet is to reduce the fat in some cases quite rapidly, and thus it answers the purpose ; but if continued too long, it becomes wearisome and injurious.

When athletes return to the usual habits of eating they sometimes rapidly increase in weight.

In their experiments in dieting, athletes have made most serious blunders. I was in London at the time of the International Boat-race between Harvard and Oxford, and ascertained from Mr. Blaikie that the American crew ate largely of vegetables, and relatively less of meat than the English crew. If the Americans had allowed their rivals to prescribe a dietary for them they could scarcely have made a worse selection. There is in my mind little question that the race was lost to the Americans on account of their unfortunate system of athletic training.

They failed not for lack of native strength, in which they were superior to their rivals ; but from lack of staying power.*

On what principle our oarsmen persisted in confining themselves to a diet which all experience has shown is unfitted to sustain amid the severest muscular exertion, in our climate at least, I never could well understand. I suspect, however, that it was inspired by some of the many popular treatises on diet by which ignorant writers in our country have wrought so much evil.

* I formed this opinion from my own observation of the appearance of the Americans near the end of the race, at Barnes' Bridge.

CHAPTER VIII.

THE objects of cookery are manifold, and cannot be completely defined.

The changes produced in food by the processes of cookery we know only in part, and therefore we cannot determine with precision the advantages and disadvantages derived from these processes.

We may, however, study at this question and may reach conclusions which, though very general, are yet in the main correct.

First of all, we must admit the premise that cookery of any kind is a luxury rather than a necessity. Food of all kinds—animal, vegetable, mineral, liquid and solid—can be and is eaten by man, as by other animals, without any preparation at all.

Those who ride on the topmost wave of civilization do not fear to eat raw beef at times, and very rare beef habitually, raw turnips, and radishes, and onions if they are so inclined, as well as the ordinary fruits.

On the other hand, savages sometimes cook what we prefer raw, for travelers say that in some tropical regions fried banana is much prized.

If we now address ourselves to the inquiry, why we

cook our food, we shall find it possible to give an answer not entirely unsatisfactory, as follows :

1. *Cooking makes our food more attractive to the senses.* Not only the appetite, but the eye and the sense of smell are gratified by the processes of baking, and roasting, and stewing, and thus the food becomes more longed for, and consequently more gratifying than it possibly could be in a raw state.

The process of cooking develops the *osmazome* on which its flavor so much depends.

2. *The process of cooking is attended or preceded by processes of cleaning, by which much in food that is repulsive and useless is cast aside.*

Fish and animals are usually dressed before they go to the gridiron or frying-pan, and it is by this process of dressing that most meats and fish are made fit food for refined humanity. Similarly, but to a less degree with fruits and vegetables, nearly all of which need more or less of lopping off, and scraping, and cleaning before they are put into the oven or stew-pan.

3. *It makes food more digestible.*

Boiling, stewing, and roasting, make flesh and fish softer and more juicy, and consequently they are more easily assimilated.

The abstract nutritive value of food is not necessarily increased with increase of its digestibility, but on the contrary, is sometimes diminished. The process of boiling especially reduces very materially the actual nutritive value of meats.

4. *The combinations made in the processes of cookery*

enhance the attractiveness, consequently the value of natural food.

Light and luscious pastry, and cakes, and puddings, creams, and sauces, and gravies, succotash, gruel and bread—these are some among many combinations by which our tables are made pleasing to the eye and to the appetite, and to the organs of digestion.

5. *Cookery* destroys the parasites with which some meats are infested.

The *trichina* of pork is sure to be destroyed if the meat be thoroughly cooked at a high temperature. Savages who live on raw meat suffer from parasites more than the civilized.

As man advances in refinement, and becomes more and more nervous and susceptible and exacting in his taste, cooking becomes more and more necessary, and dishes improve in value and increase in variety.

FRYING.—This is a method of cooking meat which has no other recommendation than convenience. It is a rude method, adapted for coarse natures, and disappears before civilization.

BOILING.—Boiling is a method of cooking meat, that is considered more economical than frying, baking or roasting. To boil meat in the best possible manner a large piece should be put into *boiling water* and kept there for a few minutes until the albumen of the surface is coagulated, so that the juice may be retained. Then cold water may be added until the temperature is somewhat relieved, and the meat cooked until it is tender.

Soups.—In making soups it is desired to extract as much nutriment of the meat as possible into the water; therefore the meat should be chopped into fine pieces and put to soak in cold water, which is gradually increased in temperature until it reaches the boiling point.

Roasting.—In roasting meat, the heat should be greatest at first, so as to *coagulate the albumen on the surface, and prevent the juice from flowing out.*

Broiling.—The method of cooking beef-steak and some varieties of fish by broiling has manifest advantages to all who have been compelled to live where steak is always fried in "spiders." The gridiron is the thermometer of civilization. In proportion as men become cultured and well-to-do, in that proportion do they give up frying and substitute broiling. In wild, ignorant and unsettled portions of our country, especially in the West and Southwest, and in distant rural districts everywhere, meat is almost always fried ; the gridiron is little known.* In broiling the heat should be very strong at *first,* and for the same reasons as in roasting.

Bread-making.—The leading principle involved in bread-making is to diffuse through the dough carbonic acid gas so as to press against the elastic gluten of the flour, and make it light and spongy. This is accom-

---

* Mr. Horace Greeley complained of this quite feelingly, as though he had suffered thereby, after his return from his late visit to Texas. He stated, furthermore, that among other civilizing influences, an army of twenty thousand good cooks were needed in that section.

plished in various ways, the best known of which is by fermentation with yeast.

Baking powders composed of *carbonate of soda* and *tartaric acid* are much used.

When mingled with the dough they effervesce like soda water, through the evolution of carbonic acid. These baking powders. when properly made are not harmful, and the bread made from them can be eaten *hot* even by those of delicate digestion. In fermented bread the process of fermentation goes on for some hours after the bread is baked, and withal fermented bread is at first moist and a little heavy.

For these two reasons fresh fermented bread is difficult of digestion, and not because it is warm.

AERATED BREAD is made by forcing into the dough the carbonic acid gas, which is generated by the action of sulphuric acid or marble dust. This process can only be satisfactorily performed on a large scale, and by the aid of powerful machinery. Bread thus made is very light and quite palatable, especially when it is toasted.

PHOSPHORATED BREAD, by Prof. Hanford's process, is made by mixing with the dough *phosphate of lime* and *bi-carbonate of soda*, so as to form neutral phosphate of lime and phosphate of soda. Bread prepared in this way contains some phosphorus, which is of undoubted advantage. The fact that the phosphorus is *not organized* is no objection, since the experience of mankind in general, of the earth-eaters in particular, and of physicians who use phosphorus, iron etc., in their practice, show conclusively that life may be sustained and health improved by the use of inorganic substances.

PASTRY.—Light crisy pastry is to be recommended, while heavy sodden crust, so-called, is to be avoided by all civilized beings.

PUDDINGS.—When puddings are light and palatable they are valuable food, and are not injurious when eaten as warm as the temperature of the body. Heavy, sodden "sailor's duff," and very hard and ugly plum puddings, are condemned alike by the palate and the stomach.

CHAPTER IX.

DIETARIES.

The object aimed at in the construction of dietaries for prisons, work-houses, and for soldiers and sailors, is to obtain the *best possible variety of nutriment for the character and duties of the class for whom it is designed, at the least possible expense.*

For those who are their own masters in all respects, and have tolerably abundant means, dietaries are useless. Let them select, of all the wide range, what suits them in quality and in quantity, and they cannot go far wrong. But for the classes I have mentioned, dietaries of some kind—good or bad—are essential; and it is the part of science to search out the lessons derived from experience, and apply them to the wants of the poor, the unfortunate, and the dependent.

It is almost needless to say, in view of the facts previously given, that the range of sustaining diet is exceedingly wide, and that any dietary based on an average of this kind must be much too little for some, and a little, or perhaps a good deal, too much for others.

Experiment shows that in prison men cannot live on one pound of bread a day; and that the poor needle-women can barely keep from dying on a diet of one and a half pounds of bread and one ounce of dripping daily.

It is found by experiment, that in prisons those who work must have more to eat than those who are idle. Dr. Edward Smith has shown that for the adult male operatives of Lancashire, two pounds and four ounces of bread sustains life ; but he regards this as a famine diet.

In regard to the quality of food for dietaries, it has been demonstrated by Playfair, Voit and Pettenkofer, that an average adult, at work, requires :

Of nitrogenous matter, 5.22 oz.
Of carbonaceous matter, 22.38 oz.

When perfectly idle, about half of this may suffice.

In regard to the proportion of the nitrogenous to the carbonaceous matter, it has been estimated that one part of the former to five or six of the latter is about correct ; but this proportion is open to great modifications, and for a limited period it is possible to work hard, with brain or with muscle, on a diet that is exclusively carbonaceous—starch, sugar, oils, and fat ; or on one that is exclusively nitrogenous, as meats. But the appetite, which knows more and better than any of us about these matters, soon calls for variety, and will have it if it can be obtained.

In the construction of dietaries, where it is not designed to punish by starvation, it is necessary to adopt a *very liberal estimate,* so that every individual shall be sure to have enough of both departments of food.

Women require one tenth less than men for the threefold reason that they weigh less, that their brains are about one tenth less than that of man, and because they do less work.

DIETARY OF BRITISH SOLDIERS IN TIME OF PEACE.

The dietary of the British soldier on home service very fairly represents the military dietaries of Europe and America. It is as follows :

Meat	12 oz.	Tea	0.16 oz.
Bread	24 oz.	Sugar	1.33 oz.
Potatoes	16 oz.	Milk	3.25 oz.
Other vegetables	8 oz.	Salt	0.25 oz.
Coffee	0.33 oz.		

Total, 65.32 oz.

A MODEL DIET.

A model diet for this climate, and for adult males of average health, should contain,—

1. About thirty-five ounces of *dry* food, composed of about one half or one third water ; three or four ounces of flesh-forming, and twelve or fifteen ounces of heat-giving material.

In the so-called heat-giving materials the proportion of carbo-hydrates to fat should be about three to one.

The amount of carbon should vary between five and ten ounces, according to the season, labor, etc. Vegetables and fruits are here included.

2. Water, to the extent of about fifty to one hundred ounces, according to season, etc.

It has been estimated that the body, on the average, needs nine parts of fat, twenty-two parts of flesh-forming substances, and sixty-nine parts of sugar and starch.

Experience shows that this proportion is not far from correct.

The intuitions of mankind have long ago discovered this fact and acted upon it.

When any article of food is deficient we supplement it by something that will compensate for these deficiencies if we can get it.

An Englishman in his own country will take about from one twenty-sixth to one twentieth of his weight in solid or liquid food; that is, from 80 to 120 ounces by weight. The proportion of solid to liquid is 1 : 2, or 1 : 1 ; but varies exceedingly.

In all solid food there is a greater or less percentage of water. Eliminating this we find that the average Englishman takes :

> Of dry food, from 22 oz. to 23 oz.
> Of water, from 60 oz. to 90 oz.

Dr. Dobell has prepared the following normal diet tables, that are, perhaps, as reliable as any. They are reliable so far as they accord with experience ; and must, of course, be greatly varied by season, temperature, occupation, etc.:

It has been estimated that the average consumption of *dry* food for Englishmen, in health, is from 700 to 750 pounds a year ; or about two pounds daily, with five or six pounds of water.

FOOD FOR TWENTY-FOUR HOURS.

Oz.	No. 1.	Oz.	No. 2.
6	Meat, poultry, or game.	18	Bread.
4	Fish.	3½	Cheese.
10	Bread.	3	Bacon.
8	Potatoes.	1½	Sugar.
2	Rice.	5	Milk.
2½	Sugar.	20	Chocolate.

$2\frac{1}{2}$ Butter.

5 Milk, (liquid).

16 Coffee, "

16 Tea, "

17 Water, "

———

89 ounces.

21 Tea.

20 Water.

———

92 ounces.

Oz. No. 3.

16 Oatmeal.

22 Milk.

$1\frac{3}{4}$ Butter.

$\frac{3}{4}$ Sugar.

49 Water.

———

$89\frac{1}{2}$ ounces.

Oz. No. 4.

25 Bread.

$3\frac{1}{4}$ Cheese.

2 Butter.

60 Water.

———

$90\frac{1}{4}$ ounces.

It will be observed that all of these tables contain all the necessary variety and quantity of food.

CHAPTER X.

THE object of adulterating food is threefold.

1. To increase its weight or size.
2. To lessen the expense of manufacture.
3. To give it attractive color, and taste.

When a dealer adds a substance to food for any of these purposes, the main object is the same—*to make money by fraud.*

BAKERS' BREAD is adulterated with potatoes, beans, rice, rye meal, and corn meal,—all of which are both harmless and nutritious,—and with alum to give it superior lightness and whiteness. Flour is also adulterated with clay, bone dust, and carbonate of magnesia.

ARROW ROOT of the first quality is adulterated with cheaper roots and starches.

MILK is chiefly diluted with water, but sometimes burnt sugar, salt, bi-carbonate of soda, gum, flour, sugar, or starch, are added.

BUTTER is adulterated with lard, water, and with a larger quantity of salt than is needed to preserve it;

and *lard* itself is in turn adulterated with water and salt.

SUGAR is adulterated for weight with lead and iron, and refined sugar by wheat flour.

The adulterations that are most injurious are those used in giving color and taste to confectionery.

Our popular candies are adulterated for bulk and weight with starch and flour—which are harmless enough—and with clay, chalk, and plaster of Paris. Among other substances used for giving color and so forth to fancy confectionery, are cochineal, Prussian blue, indigo, ultramarine, carbonates of copper and lead, cobalt, red lead, and the chromate of lead or chrome yellow, verdigris, gamboge, Brunswick green, arsenite of copper, ochre powders.

Among the condiments, *vinegar* is adulterated with water, spirits of nitre, burnt sugar, fusel oil, and acetic acid ; *ginger* with Cayenne pepper, turmeric, and corn meal ; *Cayenne pepper* with brick dust, salt, red ochre, corn meal, red lead, and Venetian red, rice and cinnabar ; *mustard* with flour, salt, and turmeric ; *cinnamon* is adulterated with *cassia*, and this in its turn with sugar and wheat flour, and arrow-root ; *curry* with lead, mercury, iron, salt, rice, and potato flour.

PICKLES. Salts of copper are sometimes added to pickles to improve their color.

It will be seen that aside from the coloring matter of confectionery, most of the adulterations of positive food, as of stimulants and narcotics, are comparatively harmless, and do more injury to the

noral sense of the trader than to the health of the
:onsumer.

POISONOUS WATER.

Probably no article of food is so frequently poisonous
,o individuals as water. Analysis of the principal drink-
ng waters shows that most of them contain poison,
hat indeed perfectly pure water, in which no poison can
)e found, is usually unpalatable. Whether the earthy
ngredients of our drinking waters exert a poisonous
nfluence depends on the individual temperament, the
abits and the state of the health. Waters which con-
ain a variety of mineral poisons are oftentimes regarded
s medicinal ; and nearly all the popular springs derive
heir popularity from the mineral substances which they
:ontain.

The chief ingredient of water that exerts a markedly
)oisonous effect in our large cities, is lead, which comes
rom the pipes ; there is little doubt that the use of lead
,ipe is always attended with more or less danger, but a
:ertain amount of lead can be habitually taken without
njury. In susceptibility to the poison of lead, individ-
als widely vary. Dr. Angus Smith says that some per-
ons are affected by one fortieth of a grain, others by
,ne tenth. Dr. Parkes regards any quantity above one
wentieth of a grain as unsafe.

Experiments have shown that the Cochituate water is
ever free from lead ; that the pipes which convey hot
,ater are more rapidly corroded than those that convey
old water, and yet "no well authenticated case of bad
,oisoning" from the Boston water has come to the
:nowledge of the State Board of Health, "although

lead pipe is almost universally used for distribution." It is not improbable, however, that many obscure cases of nervous diseases are caused by the slow action of lead taken daily into the system in the water.

Water may also acquire poisonous properties from the contents of sewers, cess-pools, and so forth, or from substances that are thrown into wells and cisterns, or from animals that fall into and are drowned in them.

CHAPTER XI.

NOT only is food liable to cause disease through excess or deficiency, or bad combination, but it may be actually diseased, and so injure more than it benefits the system.

DISEASES OF ANIMAL FOOD.

The meat of animals that have died of disease is not always or usually injurious as food, and we have already seen that putrid and decomposed flesh is eaten without apparent harm by millions of the human race. During the prevalence of the *rinderpest* in England, in 1863, "enormous quantities of meat from the diseased animals were sent to market, and sold and eaten." It has been estimated that one fifth of the meat sold in England is the meat of diseased animals. Animals that have died from pleuro-pneumonia, have been eaten without doing any harm.

On the other hand diseased sausages have been known to produce diarrhœa and debility, and Dr. Livingstone says that in Africa, the flesh of animals suffering from pleuro-pneumonia causes malignant carbuncles, even when it is thoroughly cooked. The probability is that in the process of assimilation, the injurious diseased meats, like the poison of snakes, in some way loses its injurious character, becoming an entirely different substance.

PARASITES IN MEAT.

The most familiar form of parasite in food is the *trichina spiralis*, which is found in pork. It causes terrible symptoms, the nature of which was not discovered until 1860. This worm is sometimes found in other animals besides the hog—as dogs, sheep, rats, mice, foxes, and frogs—and carnivorous birds. None of these animals seem to be worried by the presence of these parasites ; only man is a sufferer.

Other parasites are the *cysticerci* or *measles*, found in beef, and veal, and pork, which form tape worms ; and the *hydatid*, which is found in sheep and in man, and which comes from a little tapeworm. This parasite infests the dogs of Iceland, and the *hydatid* of the liver which it produces is the cause of many deaths in that country. The *flukes*, in sheep, come from the snails that they eat, and to which they are attached.

The only way to save ourselves from the visitation of these parasites is *to kill them by cooking our meat thoroughly at a high temperature*.

Meat may sometimes become unfit for food by the diet of animals. Partridges become poisonous at certain seasons in this country, by eating the laurel. Hares and prairie hens sometimes become poisonous through some food that they eat, and cattle, in some districts, obtain substances that make their flesh and their milk injurious.

DISEASED FISH.

Certain varieties of fish are always, or almost always poisonous. It is perhaps no more strange that such should be the case than that some varieties of fish are poisonous at certain seasons, or to certain individuals. That oysters, lobsters, clams, and especially mussels, may be harmful, especially in the summer season, everybody knows, and some have learned the fact by personal experience. The causes of this poisonous character of fish, are, it may be supposed, various. It may be due to the food they eat, to their not being properly cooked, or to some indefinable and mysterious peculiarity in the character of the fish that unfits it to become nutriment for a human being. One thing is certain, that it is usually impossible to distinguish poisonous fish either by their appearance or by their taste.

There is much fear of oysters and clams during the summer months, and yet only in exceptional cases do they harm any one at this season ; but that the risk is greater in warm than in cold weather, there is no doubt.

Lobsters sometimes cause cholera morbus and diarrhœa, especially when they are not thoroughly cooked, and yet they act capriciously enough, for when at public dinners several hundred partake freely of them, only a small percentage will be attacked during the following night, but those few may suffer enough for all the rest.

In the West Indies, and in other hot climes, there are fish that are always, or almost always, poisonous to almost all who eat them. Among them are the *bladder fish*, the *yellow billed sprat* and the *grey-snapper*.

When the latter is eaten by the dolphin, or congor eel, or globe-fish, it causes them also to be poisonous.

It is said the *yellow-billed sprat* is so poisonous that men have been known to die with the fish in the mouth unswallowed.

In Mauritius the parrot fishes are unwholesome at certain periods, because they eat the coral polyps.

The familiar Portuguese-man-of-war—a fish which I have seen in great numbers in the Gulf Mexico—is said to poison the fish that eat it.

Some fish are wholesome in one place and poisonous at another. Thus the *barracouta* at Trinidad can be eaten, but at Granada, an island quite near it, causes sickness.

In Havana there is a fish called *carana fallax*, which is *wholesome when small, and poisonous when large,* and therefore it is unlawful to expose for sale those that exceed two pounds three ounces in weight. It has been affirmed that the same rule applies to many other poisonous fish. It is said of the *barracouta* that when the roots of the teeth are black, it should not be eaten ; if this be true, it is the only case where there is known any definite external sign of the healthfulness or unhealthfulness of fishes.

At the period of *spawning* some fish are poisonous. It is said that there are about *seventy* species of fish that are known to be poisonous to a greater or less degree. Of these, *twenty-three* have a particularly bad reputation, and many of these are found in the Caribbean Sea.

It is true of most poisonous fish, as of poisons in general, that they act very differently with different individuals. Thus the *anchovy* of the Indian Ocean is very injurious, unless in dressing it the head and intestines are carefully and thoroughly removed ; but one half of a family that partake of it may suffer, while the other half are unaffected.

DISEASED AND POISONOUS CEREALS, ETC.

Honey is sometimes poisonous, owing to the flowers on which the bees fed. *Mould* seriously affects some constitutions when it is found on meat, fish, cheese, sausages, or bread. *Mouldy food* of all kinds should be eaten only by those who are very hungry and very poor.

Poisonous grasses sometimes get mingled with wheat, and are ground with the flour.

Ergot of rye is a famous poison that is best known in certain districts of France and Germany ; but fortunately in this country it is never heard of.

CHAPTER XII.

VERY much of the food of the world is lost through decay, and everywhere meat and fruits are much dearer than they would be if the art of preserving food for an indefinite period, without impairing its taste and nutritive value, were revealed to man. All known methods are imperfect.

FOOD PRESERVED BY DRYING.

In hot climates, fish and meat are cut into strips and exposed to the air to dry ; and the dried beef of our own country is familiar to all. Dried meat is tough, hard to digest, and not very palatable.

Dried herbs, fruits, and vegetables are abundantly known ; but no method yet devised can make dried apples or desiccated potatoes equal to fresh apples and potatoes. The life goes out with the liquid.

FOOD PRESERVED BY EXCLUDING AIR.

Air may be excluded from food by covering it with some impervious substance, as bladder ; by exhausting it from the vessel containing the food ; by heating the food in the vessel, so as to destroy the oxygen ; by filling the vessel with hot fat, or syrup, or water, and by steam. The latter method is the one that is most employed ; and food preserved in this way—as meats and canned fruits—will last for many years. It has indeed been shown that for *nearly fifty years* food preserved by this method has been kept pure.

FOOD PRESERVED BY FREEZING.

Extreme cold is a powerful antiseptic, and will preserve animal substances for years. In very cold climates, as in Norway, Siberia, and Iceland, it is customary to preserve fish and meat by freezing.

FOOD PRESERVED BY CHEMICAL SUBSTANCES THAT PREVENT DECOMPOSITION—ANTISEPTICS.

Among the agents that preserve food by their chemical action, the best known are

Common Salt ; Smoke ; Alcohol and vinegar.

The use of all these agents is so familiar that it need not be described in any detail. At best they are all imperfect methods. Salted beef is not as digestible, as palatable, or as nutritious as fresh beef ; and the same is true of fish.

Fresh herrings make good eating ; but the smoked herring—the " Cape Cod turkey,"—are poor eating for those who can get anything better ; although, like " dried beef " and smoked ham, they may, in moderate quantities, go well as a relish, or as a relief from long subsisting on unsalted food ; or, perhaps, to make us better appreciate first-class diet.

None of the salted and smoked meats or fish are as healthful as fresh meats and fish. They are less nutritive, less digestible, and more likely to cause diseases of the skin. Soldiers and sailors suffer everywhere from excess of salted and smoked food, although, as a rule, their stomachs are about as tough as the salt junk itself.

With the progress of civilization, salted and smoked food is gradually disappearing, for the reason that nervous brain-workers do not like it, nor does it like them. Among the comfortable classes of our cities where fresh meats and fish are always in the market, salt pork, or beef, or fish, (excepting codfish and salmon,) are rarely eaten, and even ham is ignored unless it be very sweet and tender.

Fifty years ago salted and smoked food was the nutriment of nearly all classes, and even now there are large sections in the country where fresh meat and fresh fish are rare luxuries.

PICKLES, of various kinds, furnish good examples of what may be done in the way of preserving food by vinegar.   They are, as the expression is, a "mere relish," stimulating the appetite for other food, but containing little nutriment themselves.

Other chemical substances used in preparing food are :

Muriate of ammonia.	Sulphurous acid fumes.
Sulphate of soda.	Acetate of ammonia.
Sulphate of potash.	Saltpetre.

The last substance is usually combined with salt in preserving meat.

It should be remarked that the preserving power of smoke is mainly due to the *creosote* contained in the oil evolved from the burning wood.

GAMGEE's PROCESS.—Prof. Gamgee has introduced a modification of an old process of preserving meat, causing the animal to breathe carbonic oxide gas until it is nearly dead. It is then killed in the ordinary manner. When dressed the animal is hung in an airtight chamber, which, after the air is exhausted, is filled with carbonic oxide gas, mingled with sulphurous acid. After being exposed to these gases for one or two days it can be hung up in dry air and will keep for months.

PREPARATIONS OF FOOD.

PEMMICAN.—Arctic voyagers have depended on pemmican, which consists of powdered meat, to which sugar and spices have been added.

LIEBIG's EXTRACT OF BEEF.—This preparation, the nutritive value of which has been considerably overestimated, is yet an excellent substitute for fresh meat : and in soup fulfills a desirable purpose.

CONDENSED MILK.—For coffee, condensed milk is preferable, in the opinion of many, to cream, or the milk uncondensed.

It is prepared by evaporating the water from the milk.

LIEBIG's FOOD FOR INFANTS.—This popular preparation consists of

| Malt. | Milk, and |
| Wheaten flour. | Bi-carbonate of potash. |

The malt changes the starch to sugar ; the bi-carbonate of potash aids this change, and neutralizes the acids.

CHAPTER XIII.

THE size or weight, the fatness or leanness, shortness or tallness, of races or of individuals depend on a variety of conditions, none of which are fully understood. Race and climate, which, in indefinite periods of time, may be regarded as the great causes of the differences of race, diet, external conditions of wealth or poverty, education, and the numberless social influences of civilization ; all these elements together find their resultant in man ; and whether he is to be light or heavy, fat as a porpoise or lean as a rail, depends on all of these factors in their combination.

Some of these factors, and the more important ones —organization and climate—are certainly beyond our control ; and even our diet and external conditions are only to a limited extent subject to our will.

That the *quantity* of food eaten does not determine the fatness or leanness of individuals is well known, and is a matter of superficial observation ; for some who are large as barrels are light eaters, while others who seem to be almost transparent are mighty performers at the table.

It is the *quality* rather than the quantity of food that makes us fat or lean. The more we eat of some varieties of food the leaner we shall become.

It appears from the researches of Gould, during the late war, that a residence in the Western States during the period of growth was favorable to stature, and that the Kentuckians are taller than the natives of some of the Eastern States has long been believed. It has been supposed that the inhabitants of regions where limestone abounds in the water become taller thereby.

It is for these reasons in their combination, I suppose, that almost everywhere, the lower orders are smaller, shorter and lighter on the average than the higher orders. This distinction will apply to both sexes and to all ages. Take at random one thousand merchants, professional men, or freeholding farmers, and they will outweigh an equal number of mechanics and laborers in this or in any other country.

The muscles of the laboring class are large and firmly knit, and they are capable of severe toil ; but it takes fewer yards to clothe them.

Much more than is credited, toil either of the brain or of the muscle works off the adipose tissue and keeps us thin, hence we often find that cadaverous, nervous and active organizations are vigorous eaters, while the phlegmatic, indifferent and indolent may be very sparing in their diet. It is not the money that we receive but what we save that makes us rich; it is not what we eat and digest but what we hoard up in the body that makes us fat.

I am told that in Norway, the rich people are almost always fat, and travelers in Siberia have remarked that adipose tissue and dollars seem to have some mysterious connection.

The only explanation that can well be put for this phenomenon is that the rich have less depressing anxiety, more and better food to eat, and more time to eat it in, and have less severe muscular labor. I am inclined to think that the difference in respect to weight between the so-called higher and lower classes is less marked in this country than in Europe, for it is certain that not a few of our wealthy and cultured citizens could be weighed on light scales, and some of the poorest of the poor are heavy and pursy, as well as muscular.

The foods that produce fat are :

1. *Sugar and sweet* substances, as sugar, molasses, honey, etc.
2. *Saccharine fruits*—sweet apples, peaches, etc.
3. *Starchy substances*—bread, potatoes, puddings, cakes, etc.
4. *Fats and oils*—fat meat, butter, milk, cream and oil.

Those who wish to grow fat should live as much as possible on these substances. A very frequent mistake is to suppose that lean meat is fattening.

Liebig, in experimenting on grease, found that a lean goose, weighing four pounds, gained five pounds in thirty-six days on maize only ; of these five pounds, three and a half were pure fat. He obtained similar results by his experiments with sugar.

It is said that in the Seraglio at Tripoli, the ladies are fattened on time by flour, honey, rich soups and frequent bathing. Fifteen days is all that is necessary to make them as plump as an ortolan.

Humboldt says the negroes and the free people of South America who work on the plantations, drink the milk of the *cow-tree* and grow sensibly fatter during the season when it is most abundant.

The Touaricks of the desert of Sahara live four months on nothing but camel's milk, and on this diet the children grow like lions.

The sisters of Kamraisi, an African chieftain, were kept confined in the palace, and fed each on the *produce of from ten to twenty cows*. They grew so fat they could not walk. It required eight men to lift one of them on a litter. The negroes of the West Indies and of the Southern States grow fat during the sugar making season.

Many persons find by experience that when they live entirely at ease, and change their diet from meats to saccharine fruits, cakes, pastry, milk, cream, etc., that up to a certain point they gain in weight quite rapidly. Often have I observed this in my own experience. During my summer vacations in the country, a daily increase in weight of from half to three quarters of a pound for a week or so never surprises me, especially when living on fat-producing food ; but the limit is soon reached, and on returning to hard work and ordinary fare, the surplus fat disappears as rapidly as it came.

Fat bears no necessary or constant relation to health, for the best of health is possible for those who are, as well as those who are not, provided with a large amount of adipose tissue.

There is, however, a point of extreme leanness on the one hand, and extreme corpulency on the other, that the body, when in perfect health, does not pass. This point varies with individuals, and, of course, with age and sex. The sweets, oils, fats, starches, which make fat are *not the best adapted for brain-workers or athletes* as an exclusive diet ; and those who adopt this fattening process must not be surprised to find that both brain and muscle lose something of their vigor in proportion as the body increases in weight.

In order for lean persons to become fat it *is necessary to reverse the process by which the fat become lean.*

1. They should live as exclusively as possible on substances that contain little starch, sugar, fat,—as lean meat and dry bread.

2. They should sleep little and work hard with brain or muscle, or with both.

By this process it is possible for the corpulent, even the extremely corpulent, to reduce their weight to an extent that will afford them great relief.

A most remarkable illustration of the efficiency of this method of treatment was afforded in the person of Mr. Banting, of England, the author of a little pamphlet on corpulence, which a few years since created such an excitement in Great Britain.

Mr. Banting was five feet five inches in height, and weighed 202 pounds. At the end of twelve months he had reduced his weight to 150 pounds.

His daily solid diet was about 3.21 ounces of nitrogenous matter and 4.06 ounces of carbonaceous matter. The average man in health and at work requires *five or six times as much carbonaceous matter,* and decidedly more of the nitrogenous.

The special articles of diet of Mr. Banting were beef, mutton, fish, bacon, dry toast and biscuit, poultry, game, tea, coffee, claret and sherry in small quantities, and a night-cap of gin, whiskey or brandy, or wine.

He *abstained* from the following articles : pork, veal, salmon, eels, herrings, sugar, milk, and all sorts of vege-

tables grown underground, and nearly all fatty and farinaceous substances. He daily drank 43 ounces of
liquids.

On this diet he kept himself for *seven* years at the
weight of 150 pounds.

He found, what other experience confirms, that *sugar
was the most powerful of all fatteners.* If he took but one
ounce a day it increased his weight a pound in seven
days.

Temperaments differ in their ability to bear this rigid
and exclusive system of dieting, and some of the fol-
lowers of Banting have regretted that he was ever
born. On the other hand, not a few have derived suggestions from his experience for which they have reason
to be grateful. The doctrine he advocates is a sound
one ; but then those who embrace it should keep firm
hold of their common sense at the same time, and study
diligently their own constitutions.

A work like that of Banting would hardly have ex-
cited a wide interest in this country, for the reason
that comparatively few Americans are overburdened
with fat. A thousand Englishmen would weigh considerably more than a thousand Americans of the same
social station, and if some American retired manufac-
turer shall succeed in making a cadaverous body fat as
Banting did in making a fat body thin, his name will
become a household word.

In this country, the first part of this chapter will
probably be of greater practical value than the last.

CHAPTER XIV.

THE diseases of the last generation usually required fasting, bleeding and purging; the chronic diseases of the present generation usually require feeding, stimulants and tonics.

The majority of the chronic invalids of our time will come under one of these three classes—the nervous, the dyspeptic and the consumptive.

The dyspepsia of our day is largely a nervous disease, and is essentially different from the indigestion of coarse nations, and among savage races, where bad food or the lack of food may produce an inflamed condition of the nervous membrane of the stomach, giving rise to symptoms of dyspepsia.

The dyspepsia of our times is usually a disease of debility, and must be treated by measures that tend to raise the tone of the system.

The nervous and dyspeptic need as generous a diet as can be digested without great pain.

The four practical rules of dietetics apply in all their force to chronic nervous invalids, whether dyspepsia be one of their symptoms or not. For the weak as for the

strong the appetite is the best guide that nature has
given us, although it may itself become diseased, especially in dyspepsia, and so deceive us.

But the *sensations following immediately, or within a
few hours after*, meals will inform us if the appetite has
been guilty of treachery, and has allowed us to take
more food than we can digest without severe distress.

A moderate amount of pain or uneasiness after meals,
especially after the principal meal of the day, dyspeptics should be willing to endure—for they have a choice
of three evils : to abstain entirely from eating and die
of starvation ; to eat less than the system requires and
thus gradually grow weaker and lose still more of the
power of digesting, or to eat as much as can be digested
without severe distress, and thus to supply the system
with the means of recovery.   Of these three evils the
last is certainly the least, and is the one which should
be chosen.

Digestive organs that have long been nervously weak
become very sensitive and capricious, and will not tole-
rate even a mouthful of stale bread or of cracker, or a tea-
spoonful of milk or water, without entering a painful
protest, and perhaps will at once proceed to eject the
intruder.   Such cases are to be managed with studied
caution, gentleness and firmness combined.   The object
kept in view should be to give the stomach as much to
do as it can bear, not without pain, for there will be
pain whether it does infinitesimally little or lies idle, but
without exhausting it beyond ready recuperation.

Those who like the taste of Graham or Indian bread,
and have stomachs strong enough to digest it, may
very properly indulge that taste as opportunity offers,
but should not allow themselves to become slaves to it,

nor to be annoyed if they are by circumstances deprived of it. The disciples of Grahamism usually forget that bread made from unbolted flour or from Indian meal, though in some respects more nutritious than bread made of fine flour, is much less easy of digestion, and to many temperaments decidedly injurious. It usually happens that the most strenuous advocates of coarse diet are those who are least able to bear it, whose stomachs rebel against it and only receive and digest it under protest. It is better for such to cast aside all questions of chemistry and eat what they like best, even at the sure risk of committing some grave errors, than to spend life in perpetual worry. It is true of diet as of exercise, that what is best enjoyed is best digested.

A cheerful spirit covers a multitude of sins. He who takes his meals joyously, even though he eat forbidden fruit, will be more edified thereby than he whose food fulfills every requirement of chemistry but is swallowed in disgust or apprehension. The life of a man who honestly strives to fulfill every jot and title of an arbitrary gospel of health, at all hazards and at absolute cost, is very apt to be exceedingly dreary. To the physical exhaustion and pain that results from insufficient nutrition is frequently added the greater agony of a worried conscience.

Dyspeptics should with all their might avoid the following errors :

1. They should not weaken the organs of digestion by giving them too little to do.

The digestive organs are largely composed of muscles, and are supplied freely with nerves, and are very intimately connected with the sympathetic nervous system,

and the brain and spinal cord, and in their exercise
they should be governed by the same laws as the muscular system in general. If, therefore, the stomach be
not supplied with food sufficient to give it needful
employment, it becomes flabby, like an arm that has
been tied in a sling or a leg that has been fastened
in a fracture box. A radical mistake with dyspeptics
is to make their weak digestive organs still weaker
by giving them too much rest. The tables of Dr.
Beaumont, showing the relative digestibility in the
stomach of the different articles of food, have wrought
much evil in various ways, especially to dyspeptics,
who are prone to select from the list those articles
that are most rapidly changed into chyme in the
stomach. Therefore men have tried to live on rice,
raw eggs, tripe and other comparatively innutritious substances, which, according to Beaumont, remained in the
stomach of Martin, on whom he experimented, only
about an hour or one hour and a half.*

A secondary and still more serious result of living on
food that is digested in a very short time, is that the
body becomes badly nourished and grows weaker, and
this weakness reacts on the digestive organs and makes
them still more incapable of performing their functions.

2. They should not allow the stomach to go long
empty.

Their motto should be to *eat little and often;* if necessary, four or five times a day ; but at no time to load
the delicate organs with a burden that they cannot bear.

---

* To generalize from a series of experiments on a *single* individual in regard
to the relative digestibility of different articles of food is manifestly unscientific, and for that reason, as well as for the reason mentioned above, I have
departed from the universal custom of writers on diet, and refrained from publishing the tables of Dr. Beaumont.

The law is that the weaker and more nervous the individual, the less able he is to bear fasting, and it is the weak and nervous who are most exhorted to fast. To go over a meal is a bad policy for the dyspeptics of this generation, however good the treatment may have been for the gluttons of the middle ages.

3. They should not confine themselves to exclusively coarse and loosening food. -

Next to not eating at all, or eating too little, is the cruel habit of living on bran bread, grits, and fruits, in order to cure the constipation that is so often associated with dyspepsia.

Constipation is a *result* of indigestion and ill health, in general, far more than it is the cause. The alimentary canal is a *barometer* indicating the condition of the system ; if the indications are bad, we should seek to right, not the barometer, but the system. If the health every way be good, and the digestion what it should be, the bowels will take care of themselves, with only a reasonable and unvarying attention to the regular performance of their functions.

A great objection to much of the coarse food that dyspeptics so often take is that it is unpalatable, which is objection enough ; but more than that, it is oftentimes so hard of digestion that it overtasks the feeble stomach, and makes the whole process of assimilation still more faulty, and so aggravates the very constipation it was designed to cure.

It is a dreary business to take food as medicine.

The notion that coarse food, grits, etc., are so very more nutritious than fine flour, which has been so industriously circulated by writers, is as untrue as it is widely believed.

4. They should not over-exercise brain or muscle so long as the organs of digestion are unable to assimilate a fair quantity of nutriment.

If the engine is old and worn out, and in danger of exploding, we must carry less steam. Gradually and with the increase of digestive power, there should be increase of muscular activity. For the nervous and delicate, especially if they are ambitious and energetic, there is more danger that they will do too much rather than too little—that they will exceed their strength rather than fall below it—that they will continually draw on their slender capital instead of wisely living within their little income and making their expenses always a little less than their profits.

Next to starving themselves, the greatest mistake of the nervously dyspeptic is to over-exercise, especially before breakfast, on an empty stomach. *Better occasionally overeat than habitually eat too little. Better do nothing than do too much.*

A UNIVERSAL BILL OF FARE.

To assign a *definite* diet for the dyspeptic would be inconsistent with the system of hygiene taught in this work. I have everywhere insisted that every individual must study his own experience. There are, however, some articles of diet, which, like some medicines, are less capricious and uncertain in their action than others and therefore more likely to suit the average constitution in health and chronic disease.

If I were forced, under penalty of my life, to indicate a short bill of fare that would be suited to all climates and seasons, to both sexes and all ages except infancy, which with least liability to overtax or undertax the organs of digestion, should diminish the delicacy of the weak and increase the strength of the strong, which should contain all the needful elements of nutrition, and be equally adapted for breakfast, dinner, tea or lunch, which should sustain the brain-worker alike with the muscle-worker, and be the longest to pall on the taste of any, I should name the following :

Tender beef steak, fat and lean, broiled.
Roast potatoes.
Bolted wheaten flour bread, stale and light.
Butter.
A cup of weak tea.

DIET OF NERVOUS INVALIDS.

The diet of nervous invalids of all kinds, should be regulated on the general principles already laid down for dyspeptics and for those in health. It should always be remembered that the majority of nervous patients are underfed ; that they need building up and not pulling down. Nervous patients suffer not from excess but from deficiency of nervous force. Almost always they have too little rather than too much blood ; feeble and capricious appetites and delicate digestion that needs to be tenderly and wisely treated, so that the system may obtain as much as possible of nutriment, without overtasking the organs of assimilation.

THE DIET OF CONSUMPTIVES.

Nearly all medical observers now agree that consumption is a disease of debility, and that it is to be relieved and cured—if relieved and cured at all—by measures calculated to raise the general tone of the system—and that among these tonic measures a nutritious diet takes the first rank.

Consumption, like nervousness, is largely though not entirely a disease of civilization, for although it is found in nearly all parts of the globe, and among the wild and uncultured races, its worst ravages are made among the best people. It is very natural to inquire whether the evolution in diet, that, as we have seen, has kept pace with the evolution of humanity, has not something to do with this increasing and fatal frequency of consumption among us,—whether it does, indeed play as important a part in producing this disease as over-toil and heavy and impure air or vicious habits.

It is one of the discoveries of recent years that the majority of consumptives dislike fat meat, cut it off from the lean when it is served out to them, and find difficulty in eating or digesting oily food of any kind. From my own researches, I should judge that at least *three out of five consumptives, and consumptively disposed persons, have a natural repugnance to fat meat.* Large families in whom the tendency to consumption is inherited, can be found, all of whose members reject not only pork, but the fatty portions of beef and mutton. It does not follow that all those who dislike fat are destined to die of consumption, but such dislike is, to say the least, not a very favorable symptom, and indicates a type of constitution that is probably not of the strongest.

The more closely I study this subject the more and more thoroughly I become convinced that *distaste for fat in adults* is both an effect and a cause of modern debility, and that it aggravates if it does not originate chronic diseases of the nervous system.

There are facts enough already given in this volume to make this view very plausible. Both in the regions of extreme heat and extreme cold, fats and oils are used with every meal, and to a degree which, at the very mention, almost sickens us ; and among nearly all wild races everywhere, and among our early ancestors in Germany and Great Britain, and even by our fathers of the past generations, fat pork and gravy were used with a freedom which very few in our large cities would now attempt to imitate.* The modern constitution can not digest fat as our ancestors could, and the distaste for it which is now so prevalent, and I believe increasing, is a sign of this inability. It is in obedience to this peculiarity of the modern temperament that pork has so generally given way to beef, sweet butter and cream to oils and gravies, the gridiron to the frying pan.

This distaste for fats, and corresponding inability to digest them, operates as a *cause* of debility by depriving the body of a means of nourishment which, in all ages and nearly all climes, has been found to be indispensable to the highest health and activity.

Bearing in mind that fats enter into the composition of brain and all nervous tissue, and that the brain-toil of modern life causes vast expenditure of the brain and nervous system—which expenditure must be supplied partially, if not entirely, by the food—we can see readily enough that debility of any kind, or of all kinds, might be the logical result of long abstinence from fatty and

---

* The Greeks and early Romans were great consumers of fat, and among them pork was held in great esteem.

oily food ; and when we consider that this distaste for necessary nutriment is transmitted from parent to child, the nervousness and debility of modern life loses very much of its mystery.

Very powerfully this reasoning is reinforced by the undeniable fact that the women of our land, who are vastly weaker than the men, are also more afraid of fat meats, and more sparing in the use of fatty food of every sort. Children, too, in their early years of debility, while the question whether they are to live or die is yet undecided, often reject fats and gravies, or eat them only under compulsion ; and any exhibition of fondness for them is, I believe, a good sign.

In children, and in adults, excess in, or a disproportionate use of, fat may cause eruptions on the skin, and very likely other bodily evils, but it does not cause nervous debility.

The remedy for this vast evil suggests itself : it is to *prepare fatty and oily food in such a way that the modern palate can relish or endure, and the modern stomach can digest it:*

Cod liver oil is a step in the right way ; and cream is becoming more popular than it ever was, and can be taken by many who never cease to dislike cod liver oil. Sweet oil, and sweet butter, and elegant gravies may be experimented with ; and the fatty portions of beef, when combined with farinaceous substances, may in turn become quite palatable. Already we have seen that the appetite, and with it the digestive capacity, can be gradually trained so that articles originally repulsive and indigestible, can become attractive and easy of assimilation.

In all our experiments in this direction, common

sense should never be laid aside. To overeat, to force down repulsive food that rises in rebellion as soon as swallowed, will only defeat our purpose. The appetite and stomach should be led, and not driven ; *educated*, and not commanded.

Consumption, as well as nervous diseases, should be taken in time, and combated by scientific and generous dieting *before* the disease develops itself ; for after the deposit in the lungs becomes fully and easily recognized, the battle is hard and the issue doubtful.

In the use of the stronger *stimulants and narcotics*, as alcoholic liquors, tobacco, and coffee, the dyspeptic, and the nervous, and the consumptive, can never be emancipated from the law which requires every man to work out his own salvation.

There are consumptives who owe their lives to a right use of negative food ; there are others now in their graves, who, if they had never tried it, might now be living.

THE DIET OF THE SICK ROOM.

The laws of dietetics are in force in disease as well as in health, and in acute as well as chronic diseases. In sickness as in health, the appetite, with all its liability to go astray, is the best guide that nature has given us. In the incipient and early stages of acute disease there is usually little desire for food, and little ability to digest it. When fever enters, appetite departs, and then nothing is gained by compelling the stomach to receive food.

When convalescence begins, appetite, and with it the power to assimilate, reappears, and with proper judgment can be studiously gratified. In sickness as in health, the system requires a variety of nutriment, which must be modified and adapted to the climate, season, age, sex, and individual temperament. In sickness, as in health, the little that is eaten should be taken, not as a medicine, but joyously and for the pleasure of it.

The *special* indications of diet for the sick must be met as they arise, and under the direction of a medical attendant.

THE END.

POPULAR SCIENCE—Mental, Moral, Political.

B ASCOM. Principles of Psychology. By John Bascom, Professor in Williams College. 12mo, pp. 350, $1.75.

"All success to the students of physical science; but each of its fields may have its triumphs, and the secrets of mind remain as unapproachable as hitherto. With philosophy and not without it, under its own laws and not under the laws of a lower realm, must be found those clues of success, those principles of investigation, which can alone place this highest form of knowledge in its true position. The following treatise is at least a patient effort to make a contribution to this, amid all failures, chief department of thought."—*Extract from Preface.*

B ASCOM. Science, Philosophy, and Religion. By John Bascom, author of Psychology, etc. 12mo, cloth, $1.75.

† B LACKWELL. Studies in General Science. By Antoinette Brown Blackwell. 12mo (uniform with Child's "Benedicite"). Cloth extra, $2.25.

"The writer evinces admirable gifts both as a student and thinker. She brings a sincere and earnest mind to the investigation of truth."—*N. Y. Tribune.*

"The idea of the work is an excellent one, and it is ably developed."—*Boston Transcript.*r

C HADBOURNE. Natural Theology; or, Nature and the Bible from the same Author. Lectures delivered before the Lowell Institute, Boston. By P. A. Chadbourne, A.M., M.D., President of University of Wisconsin. 12mo, cloth, $2.00; Student's Edition, $1.75.

"Prof. Chadbourne's book is among the few metaphysical ones now published, which, once taken up, cannot be laid aside unread. It is written in a perspicuous, animated style, combining depth of thought and grace of diction, with a total absence of ambitious display."—*Washington National Republic.*

"In diction, method, and spirit, the volume is attractive and distinctive to a rare degree."—*Boston Traveller.*

C HADBOURNE. Lectures on Instinct. By P. A. Chadbourne, author of "Natural Theology." 12mo. (*In press.*)

H YACINTHE. Life, Speeches, and Discourses of Pere Hyacinthe. Edited by Rev. L. W. Bacon. One vol. 12mo, cloth, $1.25.

"We are quite sure that these Discourses will increase Father Hyacinthe's reputation among us, as a man of rare intellectual power, genuine eloquence, ripe scholarship, and most generous sympathies."—*National Baptist, Philadelphia.*

"The Discourses will be found fully up to the high expectation formed from the great priest's protests against the trammels of Romish dogmatism."—*Rochester Democrat.*

H YACINTHE. The Family. A Series of Discourses by Father Hyacinthe. To which are added, The Education of the Working Classes; The Church—Six Conferences; Speeches and Addresses. With an Historical Introduction. By Hon. John Bigelow. 12mo, $1.50.

N. B.—Both books are published under Father Hyacinthe's sanction, and he receives a copyright on the sales.

SMITH. A Manual of Political Economy. By E. Peshine Smith. 12mo, $1.50.

*** A comprehensive text-book, specially suggested and approved by Henry C. Carey and other eminent political economists.

WHAT IS FREE TRADE? By Emile Walter. 12mo, $1.00.

"An unanswerable argument against the follies of protection, and a stinging satire on the advocates of that policy, which would enrich us by doubling our expenses. Wit and sarcasm of the sharpest and brightest sort are used by the author with great effect."—*N. Y. Citizen.*

"The most telling statements of the leading principles of the free trade theory ever published, and is, perhaps, unsurpassed in the happiness of its illustrations."—*The Nation.*

IV.—POPULAR SCIENCE.—Physiology, Health, Domestic Life.

PUTNAM'S HANDY-BOOK SERIES FOR THE FAMILY.

BEARD. Eating and Drinking: Food and Diet in Health and Disease. By Geo. M. Beard, M.D. 12mo, paper, $ cloth, $

BEARD. Stimulants and Narcotics, Medically and Morally considered. By Geo. M. Beard, M.D., 12mo, paper, $ cloth, $

GRISCOM, J. H. M.D. ON THE USE OF TOBACCO. 32mo, paper, $0.25.

HINTON. Health and its Conditions. By James Hinton, author of "Life in Nature," "Man and his Dwelling Place," &c. 12mo, $1.50.

HOPE. Till the Doctor Comes, and How to Help Him. A Manual for Emergencies, Accidents, &c. By Geo. A. Hope, M.D. Revised, with additions, by a New York Physician. 12mo, $0.30; paper, $0.60.

WHAT SHALL WE EAT? A Manual for Housekeepers. 12mo, $0.80.

The design of this Manual is to suggest what is seasonable for the table, each day in the week; and how it shall be cooked, without the trouble of thinking. It provides an agreeable variety, which may be changed to suit the income of the reader. A collection of Pickles and Sauces of rare merit forms a desirable addition at the end.

†SWEETSER. Human Life: Its Conditions and Duration. By Wm. Sweetser, M.D. 12mo, $1.50.

"The subject is curious and interesting: the reason is logical and lucid. Some of the facts are very impressive."—*Boston Transcript.*

"A sensible and well-written treatise."—*N. Y. Albion.*

WHAT MAKES ME GROW? or, Walks and Talks with Amy Dudley. With two illustrations by Frolich. 16mo, cloth extra, $1.

*** A charming and useful little book for juveniles from six to twelve years. It is well adapted also for Sunday-school libraries.

VI.—EDUCATIONAL.—Drawing and Painting.

CAVE. The Cave Method of learning to Draw from Memory. By Madame E. Cavé. From fourth Parisian edition. 12mo, cloth, $1.

"This is the *only method of drawing which really teaches anything.* In publishing the remarkable treatise, in which she unfolds, with surprising interest, the results of her observations upon the teaching of drawing, and the ingenious methods she applies, Madame Cavé . . . renders invaluable service to all who have marked out for themselves a career of Art."—*Extract from a long review in the Revue des Deux Mondes, written by Delacroix.*

"It is interesting and valuable."—D. Huntington, *Prest. Nat. Acad.*

"Should be used by every teacher of Drawing in America."—*City Item, Phila.*

"We wish that Madame Cavé had published this work half a century ago, that we might have been instructed in this enviable accomplishment."—*Harper's Magazine.*

CAVE. The Cave Method of Teaching Color. 12mo, cloth, $1.

*** This work was referred, by the French Minister of Public Instruction, to a commission of ten eminent artists and officials, whose report, written by M. Delacroix, was unanimously adopted, endorsing and approving the work. The Minister, thereupon, by a decree, authorized the use of it in the French Normal Schools.

I. The Gauzes (framed) are now ready. Price $1 each.

II. The Stand for the Gauze. Price $1.50.

III. Methode Cave, *pour apprendre à dessiner* juste et de mémoire, d'après les principes d'Albert Durer et de Leonardo da Vinci. Approved by the Minister of Public Instruction, and by Messrs. Delacroix. H. Verhet, etc. In eight series, folio, paper covers. Price $2.25 each.

N. B.—The Crayons, Paper, and other articles mentioned in the Cavé Method may be obtained of any dealer in Artists' Materials.

VII.—ORATORY AND READING.

THE STUDENTS' OWN SPEAKER. A Manual of Oratory. By Paul Reeves. 12mo, boards, $0.75; cloth, $0.90.

The "Student's Own Book," by Paul Reeves, which forms the first of the Handy-Book Series, is notable among other points in giving "a good deal for the money." The amount of matter in this book, which is in clear and neat, though small type, fully equals that in other books of twice the size and cost. It contains many new pieces not to be found in any of the school text-books. It aims to meet the wants of a large number outside of the school-room, while it is also well adapted for school use.

The *Philadelphia Inquirer* says of it :

"The general rules laid down and the suggestions thrown out are excellent, while the pieces furnished for declamation are well chosen. The book is one deserving a wide circulation."

Another good authority says :

"We have never before seen a collection so admirably adapted for its purpose. Prose and verse, humor, eloquence, description, alteration, burlesque discourse of every kind. . . . . . For schools, clubs, and fireside amusement, it will be found an almost inexhaustible source of entertainment. . . The instruction . . . . is sensible and practical."

THE CRAYON READER. Comprising Selections from the writings of Washington Irving. For Schools and Young Persons. 12mo, cloth, $1.

IRVING'S SKETCH-BOOK. Students' Edition. 16mo, cloth, $1.25.

THE BEST READING. A Guide for Librarians and Book-Buyers. Including Suggestions for Household Libraries, with Hints for Systematic Reading, and Lists of the most desirable and important Books, under 500 subject-headings. (*In press.*)

VIII.—MECHANICS.—PRACTICAL CARPENTRY.

THE YOUNG MECHANIC. Containing Directions for the use of all kinds of Tools, and for the construction of Steam Engines and Mechanical Models, including the Art of Turning in Wood and Metal. By the author of "The Lathe and its Uses," etc. Authorized reprint from the English Edition, with corrections, etc. Illustrated, small 4to, cloth extra, $1.75.

IX.—LANGUAGES.

KLIPSTEIN'S GRAMMAR OF THE ANGLO-SAXON LANGUAGE. 12mo, cloth, $1.50.

PREU'S GERMAN PRIMER. Illustrated. Square 16mo, cloth extra, $1.

"The title of German Primer suggests at once the plan of the work : Practice before Theory, without regard to the age of the Scholar."

PREU'S FIRST STEPS IN GERMAN. 8vo, $1.25.

X.—SCIENCE.

DENISON'S ASTRONOMY, WITHOUT MATHEMATICS. 12mo, cloth, $1.75.

FAY'S GREAT OUTLINE OF GEOGRAPHY. Third edition, enlarged and improved, with an Atlas of colored maps, beautifully engraved. Student's Edition. Text-Book, $0.75 ; Atlas, $2.

—— Introductory Geography. Small quarto. *In Press.*

GUILLEMIN. The Heavens : An illustrated Hand-Book of Popular Astronomy. 8vo, cloth, $4.50.

MOLLOY'S GEOLOGY AND REVELATION. Illustrated. 12mo, $2.25.

† OTIS (Calvin).   Sacred and Constructive Art.   12mo, cloth, $1.25.

ST. JOHN.   Elements of Geology.   Illustrated.   12mo, $1.50.

XI.—POLITICAL ECONOMY.

MANUAL OF POLITICAL ECONOMY.   By Prof. E. P. Smith. 12mo, cloth, $1.50.

WHAT IS FREE TRADE.   By Emile Walter.   12mo, $1.

XII.—HISTORY.

IRVING'S WASHINGTON AND THE REVOLUTION.   Student's Edition, $2.25.

IRVING'S COLUMBUS.   Student's Edition, $1.50.

LOSSING'S HISTORY OF ENGLAND.   Student's Edition, $2.50.

PUTNAM'S WORLD'S PROGRESS.   A Dictionary of Dates, &c.— Universal History.   Large 12mo, $3.50.

XIII.—KNICKERBOCKER EDITION OF STANDARD POETS.

A New and elegant LIBRARY EDITION of the most Popular POETS, carefully edited, and handsomely printed in 12mo volumes, large type, and tinted paper, and elegantly bound in new styles.

I.

THOMAS CAMPBELL'S POETICAL WORKS.   First complete edition, with a copious Life of Campbell (100 pages), by Epes Sargent. With portrait, 12mo, pp. 479, cloth extra, $2 ; extra gilt edges, $2.50 ; half calf, $4 ; morocco extra, $6.

II.

COLLINS, GRAY, AND GOLDSMITH.   Edited, with Notes, by Epes Sargent.   Complete in one volume 12mo.   Cloth extra, $2 ; extra gilt edges, $2.50 ; half calf, $4 ; morocco extra, $6.

III.

SAMUEL ROGERS' POETICAL WORKS.   Including "Italy," "Columbus," "Pleasures of Memory," etc., with portrait.   12mo, cloth extra, $2 ; extra gilt edges, $2.50 ; half calf, $4 ; morocco extra, $6.

IV.

HORACE AND JAMES SMITH'S POETICAL WORKS. Including "Rejected Addresses," etc. Complete in one vol., cloth extra, $2 extra gilt edges; $2.50; half calf, $4; morocco extra, $6.

(To be followed by Pope, Coleridge, &c.)

* This edition of the Poets will commend itself to all who desire a compact, portable, and uniform series, at once handsome, readable, and economical ; not too small for the library, and not too large for the fireside. The edition, in handsome binding, will be specially adapted for presents and school prizes.

XIV.—WORKS OF ART.—Illustrated Books.

ART, PICTORIAL AND INDUSTRIAL. An illustrated volume of Original Essays on various topics connected with Ancient and Modern Art. With fifty-eight curious and valuable illustrations from original works, produced by the autotype process. Complete in one elegant folio volume, cloth extra, gilt edges, $16.

—— The same, half morocco, gilt edges, $18.

* This work is continued monthly. Price $1.25 per number, or $13.50 per annum.

†**B**RYANT HOMESTEAD BOOK. Illustrated by J. A. Hows. Small folio, cloth extra, $6; morocco extra, $10.

FARRAGUT'S COURTS OF EUROPE. With forty illustrations by Thomas Nast, Perkins, and Warren. Small folio, cloth extra, $7.

GALLERY OF LANDSCAPE PAINTERS. Illustrations of American Scenery. From paintings by W. H. Beard, G. S. Brown, Casilear, Colman, De Haas, Gignoux, Wm. Hart, J. M. Hart, Thos. Hill, Hubbard, Inness, Kensett, Momberger, J. D. Smillie, G. H. Smillie, Whittredge. With letter-press descriptions. In one large vol. quarto, twenty-four superb engravings on steel, cloth extra bevelled, $15 ; morocco extra, $30.

* This beautiful volume is much the finest of its kind yet produced in the United States. As a gift-book of high character and solid elegance it is not surpassed, if equaled, by any European work of its price.

HOOD'S POEMS. Illustrated (Wanstead) edition. The Poems of Thomas Hood. Artists' edition. With twenty-eight illustrations by Darley Eytinge, Gustave Doré, Seccombe, Birket, Foster, and the Etching Club. The letter-press handsomely printed on tinted paper, with red lines. Small quarto, uniform with the Artists' edition of the Sketch Book. Cloth extra, with handsome stamp, gilt edges, $7.50; morocco extra, $12.50.

IRVING'S SKETCH BOOK. The Artists' edition. With 130 beautiful engraving from original designs by Huntington, Gray, Lentre, Hart, Bellows, Darley, and other eminent artists. Small quarto, cloth extra, $10; morocco extra, $16; levant morocco, $18.

—— **Christmas in England.** With twenty-one very fine illustrations from